complex housi

Complex Housing introduces an architectural type called complex housing, common to the Netherlands and found in other Northern European countries. Eight fully illustrated case studies show successful approaches to designing for density, which reflect values such as long-term planning, a right to housing, and access to light and air. The case studies demonstrate a wide range of applications including a mixture of urban and suburban sites, various numbers of dwelling units, low- to high-density approaches, different architectural styles, and organizational strategies that can be adopted in projects elsewhere. More than 350 color images.

Julia Williams Robinson, PhD, FAIA, is Professor of Architecture at the University of Minnesota, a registered architect, and author of articles and books. She has become familiar with Dutch housing through many short and long visits to the Netherlands over the past twenty years, during which time she completed a doctoral degree at Delft University of Technology. Educated in architecture and also anthropology, her work focuses on cultural aspects of architecture, on design theory, and on the discipline of architecture.

complex housing

designing for density

Julia Williams Robinson

Routledge
Taylor & Francis Group
NEW YORK AND LONDON

First published 2018
by Routledge
711 Third Avenue, New York, NY 10017

and by Routledge
2 Park Square, Milton Park, Abingdon, Oxon OX14 4RN

Routledge is an imprint of the Taylor & Francis Group, an informa business

Library of Congress Cataloguing-in-Publication Data
Names: Robinson, Julia W., author.
Title: Complex housing : designing for density / Julia Robinson.
Description: New York : Routledge, 2018. | Includes bibliographical references and index.
Identifiers: LCCN 2017012682 | ISBN 9781138192492 (hardback) |
ISBN 9781138192508 (pbk.) | ISBN 9781315639925 (ebook master) |
ISBN 9781317275473 (mobi/kindle)
Subjects: LCSH: Architecture, Domestic—Netherlands—Randstad—History—20th century. |
Architecture, Domestic—Netherlands—Randstad—History—20th century. | Land use—Netherlands—Randstad. | Population density—Netherlands—Randstad.
Classification: LCC NA7361 .R62 2018 | DDC 728.09492—dc23
LC record available at https://lccn.loc.gov/2017012682

ISBN: 978-1-138-19249-2 (hbk)
ISBN: 978-1-138-19250-8 (pbk)
ISBN: 978-1-315-63992-5 (ebk)

Typeset in Optima and Grotesk
by Apex CoVantage, LLC

This book is dedicated to the memory of
Richard Stolzenburg
whose love of the Netherlands
and
devotion to Dutch urbanism and architecture
profoundly influenced this book

Foreword

Although Julia Williams Robinson's book is written for an international audience, it is a fascinating read for someone from the country which she has researched for many years. To say that I have learned a lot from it is an understatement.

Robinson has methodically analyzed eight housing projects which meet her definition of being complex, dense and diverse. Her choice of projects is highly original. It is an inclusive selection that comprises a few projects that have received a substantial bit of media coverage, nationally and internationally, such as MVRDV's Silodam and to a lesser extent La Grande Cour, both in Amsterdam. But most of the projects in this book are certainly less publicized, and here Robinson shows a very authentic perspective on the recent production of housing projects in the Netherlands. I cannot imagine that any Dutch person would, or could, have made this selection, simply because insiders are inevitably confined to their insiders' view and the implicit biases and blind spots related to that. So reading Robinson's take on the housing architecture has been an eye-opening experience, allowing this reader to see the qualities of Lucien Kroll's transformative housing in Dordrecht, which has never been fully appreciated or credited in the Netherlands. In a similar vein, the inclusion of a sensible and sensitive project of SCALA, a proponent of contemporary traditionalism, is laudable as well. The Netherlands has always had an uneasy relationship with anything deemed unmodern, and it is refreshing to see Robinson's unbiased appreciation for the architecture of SCALA's Mieke Bosse and Peter Drijver, who are in certain circles considered to be mavericks for the paradoxical reason of adhering to the traditions and conventions of architecture.

Another value of this book is that Robinson has spoken to almost everybody involved in these eight case studies, and almost everybody else of importance in the field of housing in the Netherlands, which has enabled her to write a most insightful book on contemporary Dutch housing.

The best Dutch housing is the collective product of, in no particular order, developers, clients, architects, contractors, and municipalities, which usually appear to agree on the idea that every project is the shared responsibility of all parties, with the final product more important than the interest of any individual party. This is often reflected in the implicit and sometimes even explicit acknowledgment that making money isn't the first and foremost reason to build. Robinson offers a striking example in one of her case studies, quoting the developer who matter-of-factly mentions a significant financial loss in the same sentence he calls the project as a huge success.

From this book one may get the impression that for all parties involved in Dutch housing it is self-evident that aiming for architectural quality and providing good housing for everybody are two sides of the same coin, minted to create common value. Robinson convincingly shows that this is true for the eight selected projects, but as a Dutchman I know that there is evidently a lot of housing in the Netherlands that does not make the cut, particularly since in recent years more and more is allegedly left to the market—a crude simplification of the working of societies and economies. But at the same time, it is intriguing to see that even with a novel dominance of market parties, the importance attached to architectural quality remains somehow deeply embedded in Dutch culture.

As a European I strongly believe that the Netherlands is not unique in this respect. The idea of architecture, and housing architecture in particular, as a common good and responsibility is equally present elsewhere in Europe, from the Baltic States to Portugal, from Norway to Slovenia. And in addition to this, as a recent Canadian transplant I notice that similar attitudes towards the public, collective, societal role and importance of architecture are much harder to find in any North American context, where architecture remains an exception to the rule, and where collective housing, also dismissively designated as "projects," have a reputation that is incomparably lower than anywhere in Europe.

Given the difference between complex housing in the Netherlands and the lack thereof in the United States, it is hard to deny that there might be some change-resistant hardwiring of societies. It is as unlikely that the Dutch will ever give up the idea that architecture is a common good as it is to imagine that North Americans will ever wholeheartedly embrace this idea. But just as Robinson's book has offered me fresh insights in an architecture I assumed I was already quite familiar with and forced me to think again, I hope and expect it will enable a readership outside the Netherlands to reconsider their assumptions of what architecture of collective housing is and could be.

Hans Ibelings
Lecturer, John H. Daniels Faculty of Architecture, Landscape and Design, University of Toronto
Editor and Publisher, *The Architecture Observer*, Montreal/Amsterdam

Julia Williams Robinson in Leidsche Rijn at Scherf13
Photo by Mieke Bosse, SCALA Architecten

Preface—Origins of the Book

When I first came to the Netherlands I was astonished at the bold design of housing. Faced with a wide variety of forms and a very different approach to spatial arrangement than that found in the United States, it took a while to make sense of what I was seeing. But having been guided through the Netherlands by my late husband Richard Stolzenburg and educated by many other Dutch colleagues as well, gradually my fascination went beyond the architecture itself to include the process by which housing is planned, designed, and built. This book is a way of exploring and sharing with a wider audience some of what I have learned about Dutch housing and urbanism since the late 1990s when I began to visit the Netherlands regularly.

Perhaps of primary significance to an outsider are those aspects of housing that the Dutch take for granted, but that in other countries are not necessarily fundamental conditions: (1) the right to housing, (2) the integration of housing with settlement design (housing as urban fabric), (3) the assumption that ecology must be taken into account, (4) the commitment to using professional designers at all scales, (5), the assumption that housing will be built to last for at least 100 years, (6) the assumption that dense housing is normative, and (7) the premise that citizens needs are central and that they need to be involved in the design of the places where they live. How and why these attitudes framed Dutch housing between 1990 and 2010, the VINEX period of planning, is an underlying theme of the book.

The most fundamental idea in this regard is the Dutch government's assumption that citizens of the Netherlands have a right to be housed, and in the 1990s that the national government, and more recently the regional and local government have a responsibility to make sure its people have housing. The second fundamental idea is that housing is seen as integral with settlement design, and that the wise design of settlements is a government responsibility. Thus, when land was developed during the VINEX period, not just the developer and the local municipality were involved, but the national and regional governments and design professionals as well. At that time the specific housing project was integral to national plans for land use, water, and transportation, as well as other housing projects. In the past, the national government was even involved in financing housing, although this has changed (and may or may not affect the quality of planning of housing).

The imperative to include water in all planning decisions due to the Netherlands' situation below sea level has meant that an ecological orientation was never controversial, but embedded in the normative approach. Similarly, because of the density of the population, especially in the western part of the Netherlands, dense housing is seen as acceptable and livable. Also, because design is understood to be a central activity (more on this later), the Netherlands values its design professionals and since the early 20th century legislated participation by architects and urban designers in the planning and development of settlements and housing. Furthermore, for reasons that will be explored in greater depth, the Dutch are committed to long-term planning and see construction of the physical world as requiring an investment in attractive and durable materials and processes.

Finally, as a nation that asserts it was the first republic in the world, not only are citizens involved in the design of physical environments at several levels, but design professionals have an unusual sensitivity and commitment to the needs of inhabitants, typically providing environments with high quality of such amenities as light, ventilation, outdoor space, landscape design, building systems, and construction materials.

These issues are especially poignant because of the recent economic crisis. In the United States the situation for housing is at a crossroads. Homelessness has become a severe crisis. The period of excess that led to the recession of 2008 not only involved construction of dwellings that were overly large and very expensive, but also encouraged people to buy houses beyond their means. Subsequent to the economic disaster, many purchasers' houses were foreclosed upon with residents evicted, or houses were simply abandoned to the bank, leaving thousands of houses to stand vacant and often be vandalized to the point of destruction. As a result, the United States has been in the peculiar position of having increasing homelessness at the same time as thousands of once inhabited houses stand empty and on the market, many of which in the meantime became uninhabitable. Something is very wrong with the provision of housing in the United States. The Dutch approach is being considered against this background.

Ironically the Dutch, having seen limitations of their top-down system to precisely match the housing market, have turned away from what at first seemed to an outsider to be well-functioning national government involvement in housing, toward a primarily market-oriented system. This book sees an important dialogue between the Dutch approach of the Netherlands and other countries, specifically the American approach as represented by the United States. Although this discussion is purposely limited to a detailed discussion of how a certain type of dense housing in the Netherlands, here called complex housing, came to be, it is anticipated that such specificity may inform an international audience of its applicability in other national contexts.

Acknowledgments

This book would not have been possible without the support of family, friends, and colleagues from the Netherlands and the United States too many to name. Of special note from the Netherlands are Richard Stolzenburg, Hein de Haan, Mieke Bosse, Peter Drijver, and Ria van Dijk, whose contributions have been fundamental and enduring.

It is important to acknowledge the contribution of Hans Ibelings, who has kindly written the foreword for this book.

From the United States, Anthony Holmes was an indispensable contributor, generously giving of his time to prepare the photographs for publication.

Also essential to the book are others who have also given the gift of their time and ideas in interviews and discussions related to the book and/or to the student trips that generated it: Han van den Born, Dag Boutsen, Frits van Dongen, Lucien Kroll, Pascal Moon, Noud te Riele, Jeroen van Schooten, Roelf Steenhuis, Frans de Witte, Eric Amory, Gerard Comello, Rob Hagens, Henk Jan Hollander, Wicher Mol, Rein Willems, David Wolterbeek, Jelte Boeijenga, Koen Crabbendam, Wim de Hoop, Chris Jagtman, Ilta van de Mast, Mirjana Milanović, Ton Schaap, Philip Spangenberg, Johan Vlug, Koos and Gonny van der Elst, Olga Kahna, Gerard and Wil Onder de Linden, Annelie Steegers, Umberto Barbieri, Carel van Brugge, Jan-Wouter Bruggenkamp, Cees Doevendans, Gerrit van Es, Eek Hackfoort, Jan Frans den Haartog, Peter Hoesberger, Hans Ibelings, Wim van der Knapp, Jos Lambrechtsen, Harry de Loor, Hetty Beuger de Loor, Thijs Meier, Cees Nagelkerke, Eric Nagengast, Lisbeth van der Pol, Stephen Read, Henk van Schagen, Hans Smits, Jurriaan van Stigt, Gert Urhahn, Jacob de Visser, Paul de Vroom, Willemijn Wilms Floet, Edward Zevenbergen, and Rene van Zuuk.

A large motivating force came from the approximately one hundred students and faculty members from the University of Minnesota who participated in the trips to the Netherlands between 1999 and 2008, and inspired me to share their enthusiasm about and appreciation of Dutch housing and urbanism.

Architectural firms and developers of these projects in the Netherlands have been very kind to provide research material such as architectural drawings, articles, and real estate brochures: Atelier Pro Architekten, Atelier d'Architecture, d'Urbanism et d'Informatique (AAUI), Bouwfonds Property Development (BPD), Bouwfonds (Raboburt), de Architekten Cie, Woonstichting De Key-De Principaal, KCAP Architects&Planners, MVSA Architects, NL-Development-BV, SCALA Architecten, Steenhuis Bukman Architecten, Stichting Woonbron, and Van Dongen-Koschuch Architects & Planners.

I am grateful to those who have donated professional images: the Stadsarchief Amsterdam, Jeroen Musch, Karel Tomei, Hans Brons, Jannes Linders, Noud te Riele, Peter de Ruig, Karel Tomei, Dirk Verwoerd, Daria Scagliola, Stijn Brakkee, Steenhuis Bukman Architects, SCALA Architects, MVSA, and KCAP Architects&Planners. I also thank those former students who have shared images with me, among others: Joseph Messier, Shengyin Xu, Cynthia Lapp, and Jarrett Anderson.

The drawings in this book would not have been possible to create without my research team, whose membership of research assistants has varied over time: Shengyin Xu, Joseph Messier, Hans-Christian Karlberg, Erin Lilli, Austin Lukes, Sean Meyers, Thomas Ducastel, Jianing Song, Kristin Ehrhardt, Andra Zerbe, Jennifer Asp, Yujing Su, Corinne Deger, Jinguang Xie, Hayden Rensch, Brittany Klingler, Mandana Motamed, Margo Fredricks, Terri (Chung Wan) Leung, Tad Neeser,

Yun Koo, Max Ouellette-Howitz, Nicole Kiel, Anthony Alan Rodriguez, and Goldielyn Lopez.

Many University of Minnesota colleagues have helped with this project—too many to name, but especially supportive have been Tom Fisher, Becky Yust, Marc Swackhamer, Cynthia Jara, Joon Mornes, J. Stephen Weeks, Gail Dubrow, and Katherine Solomonson.

The University of Minnesota has been very generous in its economic support of this project. This research project would not have been possible without funding from the School of Architecture, the College of Design, the International Travel Grant, the Undergraduate Research Opportunity Program (UROP), the Imagine Fund, and the Grant-In-Aid Program (Office of the Vice President for Research).

Introduction

Housing as urban fabric: the Colorwijk neighborhood in Almere © RoVorm

Canals, ditches and windmills © Karel Tomei

Buiten Aa marine channel at Groningen shows the potential for claiming new land from the sea © Karel Tomei

The Dutch have been pumping water since the Middle Ages.

Stolzenburg 1999–2004[1]

It is amazing to realize that the Dutch government, in a country with almost 17 million people, has since the Housing Act of 1901 supported the construction of over 7.5 million housing units.[2] Until 2005, every five years or so, the national legislature of the Netherlands in conjunction with the Ministry of Housing developed a spatial plan for the nation called the Nota Ruimte, which specified overall national goals for construction and land conservation, and designated the number of housing units to be constructed in a given period. The document also designated the income level to be served as well as the localities in which the housing was to be constructed. For example, in 2003 the Nota Ruimte proposed building between 360,000 and 440,000 dwellings in the 20 years between 2010 and 2030.[3] In this chapter, we explore what we can learn from the approach to housing taken in the Netherlands, and what may be particularly Dutch that may make direct implementation difficult in other places. To people accustomed to a completely private market-driven system, the Dutch approach must be explained starting with basic cultural assumptions about the importance of planning.

Water Management in the Netherlands[4]

The attention to planning in the Netherlands has a long history based upon the survival of the nation. Since the Middle Ages, significant portions of land were claimed from the sea and inland lakes and wetlands, giving rise to the adage "God created the world, but the Dutch created the Netherlands." Thus a large part of the Netherlands has been and is below sea level, with a system of dikes, drainage canals and windmills developed to pump the water out ("Netherlands" means low-lands).[5] Groups of fields surrounded by a dike comprise a polder. Each farmer within a polder who pumped water to keep his own fields dry was dependent upon the others in his area to do the same, which led to the creation of localized Water Boards to supervise the process.

As each polder area was similarly dependent upon the adjacent polder areas, by necessity a strong interdependent work and social network was created, and the necessity for large-scale water planning

The Netherlands with and without pumping the water

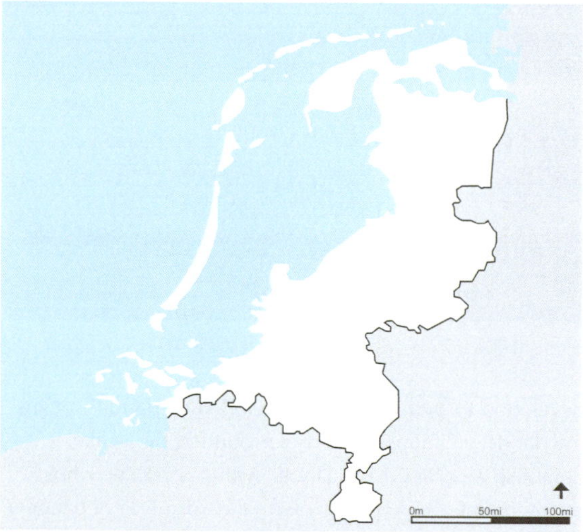

eventually led to the development of the Rijkswaterstaat, a national water planning organization (organized in 1798 as the Bureau voor den Waterstaat). This, along with the subsequent shortage of land in the western Netherlands beginning in the mid-20th century, is the basis for the strong system of planning and—according to Dutch lore—the strong community orientation and cooperative spirit of the Dutch culture.

Today planning and cooperation is even more important, as rising seas due to global warming threaten the Netherlands. Much of the western part of the country would be inundated if the Dutch ceased to maintain their system of ditches, dikes, and pumping.

Housing and Religion*

Another strong influence on Dutch culture is the Christian ethic. From the Middle Ages to the mid-20th century, religion played a very large role in daily life. Although today religion is much less practiced than in the past, there remains a very strong ethical orientation in the culture, an attitude of concern for others, and willingness to put society before oneself. So, for example, the populace is willing to pay taxes to support nationalized education and nationalized health. Everyone benefits, values, and relies upon the government support.

In the Middle Ages, the strong religious beliefs were reflected in the development of housing by wealthy families for local people who were in need, largely widows or widowers. These almshouses were typically built in a row around or along a courtyard or hof. The inhabitants shared a well—later a pump—and the courtyard provided a garden to grow food. This form of communal dwelling laid the basis for later types of multi-family housing built in blocks around open space. The scale of almshouses varied from small hofjes such as Sint Annahof in Leiden, established in the late 15th century with approximately a dozen dwellings, to large developments such as the Hofje van Nieuwkoop with sixty-two housing units, built in 1658 in The Hague by architect Pieter Post. These examples of privately

Sint Annahof, Leiden, a late 15th-century example of social housing

Hofje van Nieuwkoop, The Hague, social housing built in 1658

sponsored housing also set the stage for the later development of housing corporations.

Beginning with the late Middle Ages, the Dutch provinces experienced a great deal of religious strife among various Protestant groups. At the same time that Catholicism retained its strength in the southern part of the country, in the north and east a variety of different groups held sway. When the provinces were under Catholic Spanish rule in the 16th century, religious groups were persecuted. In reaction, after liberation from Spain in the late 16th century, the Dutch became known for religious tolerance, and many Jews and Protestants from other countries migrated to the Netherlands. Amsterdam especially was inhabited by large numbers of religious refugees. The attitude of tolerance allowed different groups to coexist, often in the same cities and towns.

By the 19th century three religious communities or "pillars" developed: Catholic, Calvinist Protestant, and neutral (which later became socialist and included many of the Jewish population that had come to the Netherlands to escape persecution in the 18th century). The pillarization allowed the different religious groups to exist side by side in Dutch society, each having its own shops, schools, professionals, and so forth in three worlds that overlapped in civic and national affairs. Pillarization continued into the mid-20th century, ending in the 1970s when the Netherlands became secularized.

The Housing Act of 1901: Housing Corporations

Up until the late 19th century, other than the almshouses, most urban housing was built by individuals within an existing pattern of canals, streets, and blocks. Cities played a coordinating role, constructing, controlling, and maintaining infrastructures and controlling new building. With industrialization came attention to housing for workers, first in rural areas, and then with mass population movement from farms to the cities. Migration to the cities created severe crowding. In the city of Amsterdam the city population increased 240%,[6] while the city boundary did not expand. Families were living together in single rooms; people were occupying windowless basement areas formerly only used for storage; housing was constructed in courtyards and alleys behind existing houses. The crowded conditions fostered the spread of cholera epidemics within the city.[7,8]

Façade, Het Schip, Eigenhaard, Amsterdam

Main room in an apartment at Het Schip with gas lighting and heating in the corner

Seeing an opportunity in the congested situation, carpenters in Amsterdam, supported by banks as investing partners, built new blocks of multi-story apartments for people of all income levels; many of these apartments were poorly constructed. Responding to this speculative housing and to the health situation, the city government created rules and ordinances that promoted health, requiring the construction of robust housing units with plumbing, toilets, and windows. Furthermore, the city itself sponsored projects for low-income residents. The city also took control over the developers with urban plans that delimited where housing could be located. Officials hired well-known and respected architects, such as H. P. Berlage, to create and implement large-scale urban designs that set a clear direction for future construction.

In 1901 the national government took formal action, passing the Housing Act that, among other things, required municipal governments to develop city plans and to regulate housing. The bill also removed housing for low-income people from the private market, providing financial support from the government, which lasted up until 2008. Along with the requirement for city planning by architects, the document required that architects design all buildings, giving a significant social role to the newly emerging architectural profession.

The government funds to support affordable housing that were a part of the Housing Act were delegated to the religious pillars to create housing corporations. These would work with architects to design and construct housing for their constituents. The government and the members of the pillars saw this as a way to emancipate the workers in each group by bringing them out of poverty into the middle class. In addition to the provision of housing, the process involved socialization (education by social workers; provision of amenities such as public schools, libraries, and public bathing and laundry facilities).

Kitchen in an apartment at Het Schip, designed for hygiene complete with gas stove (far right)

Many of the early 20th-century projects by such architects as Berlage and de Klerk, considered landmark housing today, played an important role in the upward mobility of these groups, so that their children became educated members of society. The housing corporations succeeded in providing attractive, well-constructed, affordable rental housing for Dutch citizens up to today, when many have been renovated to meet contemporary needs. At the time of construction these projects embodied the ideal middle class way to live, with bedrooms for family members, hygienic kitchens, rooms for family congregation, and such amenities as lighting, heating, and plumbing. Since then the publicly supported rental housing has been accepted as a normal way to live, so that, unlike many other countries, there is no stigma attached to living in what is called social housing. This may be due to rules that in the past have allowed inhabitants whose income surpassed the income limit for subsidy to remain in situ, paying a market rate.

Until the 1980s social housing and other private rental housing constituted 60% of the housing stock. Today, there is more emphasis on ownership, and about 30% of the housing is rental. The housing corporations survive, albeit in changed form, and are sound money-making ventures. They continue to play a significant role in the development of housing, since 2005 working in tandem with private developers to make housing with mixed income levels. The projects presented in this book reflect such a mix of income level.

Homeowners' and Tenants' Associations

In the Netherlands, since 1995 housing corporations have been encouraged to sell housing units they own within social housing projects they have developed and managed. In many cases they also retain rental units for social housing, creating mixed-tenure housing projects. To provide shared ownership of a complex when dwellings within a particular building or project are sold, Dutch law requires the establishment of a homeowners' association or Vereniging van Eigenaren (VvE), whereby owner-residents are required to participate in the maintenance of the building.[9] When a project has a mix of units owned by a housing corporation and those owned by individuals, the responsibility for maintenance is thus shared, with the tenants' interest represented by the housing corporation. Additionally, in many housing projects with rentals there is a tenants' organization that works with the housing corporation to communicate tenants' needs and desires. This enables the functioning of mixed-tenure projects.

Housing as Urban Design and Urban Fabric*

In the late 19th and early 20th centuries the architect and urban planner H. P. Berlage, employing the theories and examples of Camillo Sitte,[10] demonstrated the power

Berlage's 1915 drawing for his urban plan for Amsterdam South: "Bird's Eye View from the Bridge on the Amstel"
© Stadsarchief Amsterdam

of housing to create urban form in Amsterdam with both his extension plans and his architectural design. No doubt inspired by his work, the Housing Act of 1901 required that architects be involved in urban design and architecture. Seeing housing as integral with urban form remains fundamental to the way residential areas are designed. In stark contrast to the United States, where individual buildings (not just for housing) are typically located on singular lots, in the Netherlands, by law there is always an urban scheme that a design fits into, and most often houses are built as a group that forms a block, row, or street.

But this approach is not a radical innovation. Rather it evolved from the historic pattern of cities. Whereas in Dutch villages housing tended to evolve in rows along

the road, in cities such as Amsterdam, Haarlem, Utrecht, Dordrecht, and Den Haag housing developed in blocks.[11] As land within city walls became scarce, individual houses built along canals and streets perpendicular to canals were packed in side by side with gardens behind, creating a block of buildings surrounding an open space or hof. If cities became overcrowded, or if interior hof spaces were unusually large, buildings were constructed in the interiors of the blocks as well. Thus the ordinary house became understood as a piece of the urban fabric, part of the city wall rather than having an identity separate from other houses. So from the Middle Ages forward, when a wealthy patron decided to make a group of almshouses, or when an urban designer plans housing to sell, these have been conceived as part of the existing block fabric.

In the 19th century, developers in Amsterdam who purchased land to invest in large areas of housing also typically followed this pattern of placing housing in blocks. When cities responded to the Housing Act of 1901, they too developed urban plans conceived as neighborhoods composed of housing in blocks. Berlage's urban designs were conceived as blocks of housing that formed the streets. The urban designer set up the general urban structure, and in Berlage's case he set out the types of housing that were to be in each block, and in some cases even the façade designs. The architect was expected to develop plans that filled in the urban structure as Berlage had lain out.

However, later in the 20th century, Modern Movement urban designers challenged the notion of enclosed blocks, and created open rows of buildings that did not enclose space, such as in Rotterdam's 1940s Groenoord District in Schiedam or the Poptahof area of

The aerial view of the Colorwijk neighborhood in Almere shows a contemporary suburban conception of housing as an urban fabric © RoVorm

Wijsmullerstraat exemplifies Berlage's concept of urban fabric. Photo by Peter Drijver, SCALA Architecten

The Eco neighborhood in Almere has a distinctly suburban feel

Population density in the Netherlands by province

The Green Heart and Green Fingers of the Netherlands

Delft, built in the 1950s.[12] In these patterns, the individual garden was lost to shared public space. Although the form of the block was different from the past, being no longer closed, nevertheless these buildings were not conceived as single slabs of apartments, but as a part of the urban pattern. The larger scale and openness of the blocks was a distinct rupture with previous urban patterns.

The generation of larger urban patterns with rows of housing perhaps culminated in urban designs such as in Bijlmermeer, where the scale of the open areas between buildings became so gigantic that the pattern no longer cohered as perceptible urban fabric. In the late 20th century, in rejection of the large multi-family blocks, more emphasis was placed on family housing and home ownership, resulting in Dutch modern developments being more suburban—as exemplified in the neighborhoods of Almere of the 1980s and 1990s—despite the commitment to the conceptual link between house and city on the part of designers.

The Green Heart and Green Fingers

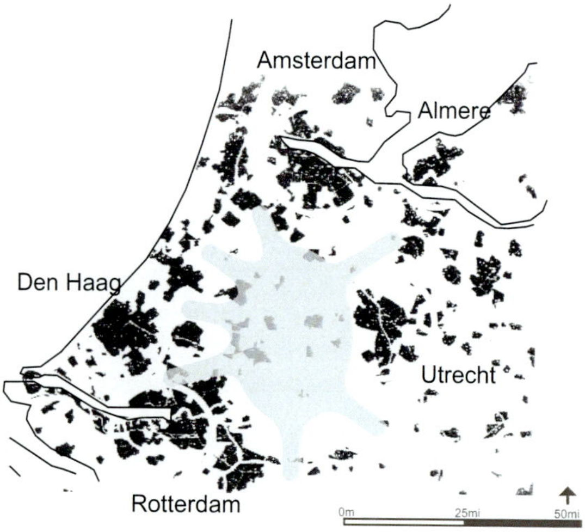

According to statistics from 2008,[13] the Netherlands is by far the densest European country, with a density of 976 people per square mile, as compared to 621 for the United Kingdom, 610 for Germany, 276 for France, and 51 for Sweden. (For a map showing the density of the Netherlands by province, See population density map.)[14] At the same time, the Netherlands has powerful links to its agricultural roots. The Dutch see their cities as being surrounded by agricultural and wooded landscapes, valuing direct access by foot and by bicycle as well as visual access when driving or traveling by train and bus.

In the densely populated Randstad area, the urbanized western area of the Netherlands that encompasses Amsterdam, The Hague, Rotterdam, and Utrecht, the pressure to develop housing conflicts with the desire to conserve green space. The planning approach has been to conserve what is called the Green Heart of the Randstad area, and rather than to have urban areas with a hard edge, to create slivers of open space that reach close to the city centers, called Green Fingers.

National Housing Policy

National housing policy for housing and communities in the Netherlands has evolved over time. According to a 2003 Aedes Vereniging van Woningcorporaties (Association of Housing Corporations) report, from 1960 to 1980 planning focused on building outside the Randstad, with the idea that people would move to communities where the housing was located.[15] But by the 1990s it was clear that this generated unforeseen problems. The new housing, designed for families, caused a migration of younger middle class people from the cities, leaving cities with a population of older affluent residents combined with those less affluent. Furthermore, since despite planning efforts most jobs continued to be located in the west of the country, creating housing in the east significantly increased automobile commuter traffic.

As a result, in the 1990s a new policy supported concentrating new housing around existing centers, while leaving the countryside as intact as possible.[16] To preserve the Green Heart of the Netherlands, the policy called for more compact, denser housing in the Randstad. To provide more income balance in suburban and urbanized areas, policies encouraged housing options for upper income people outside urban areas. Ecological considerations also have had an important place in housing policy, including maintaining sufficient housing density to support public transportation, maintaining the "green fingers" policy that supports accessibility to the countryside, and continuing traditional concern with management of runoff and other water systems. Finally, the policies attempted to reduce the use of the automobile by developing systems for pedestrian, bicycle, and public transportation systems that allowed people to move easily without having to depend upon the car.

When I visited the Netherlands in the 1990s and early 2000s, it seemed as if the policies established by Berlage at the turn of the 20th century, while modified over time to accommodate new ideas and technologies, represented Dutch practice that was immutable to change. The national government set planning and housing policy; the regional governments responded by developing appropriate water, utilities, and transportation infrastructure; and the municipalities implemented the national policy by constructing housing in response to the national agenda indefinitely.

From an American perspective, this rational approach seemed ideal, as in the United States national planning for housing had been largely abandoned. Most states had eliminated their planning agencies and the city planning that remained was primarily focused on economic development in support of commerce and job growth. Homelessness and need for affordable housing was not being effectively addressed. In contrast, the Dutch comprehensive approach to planning seemed the perfect solution. From the Dutch perspective in the 2000s, however, the housing that was being built was not always the right kind of housing in the right place. In the 1970s and 1980s, when there was an extreme housing shortage, all housing that was built was rented or sold, whereas in the 1990s some housing remained unoccupied. The American approach, where developers rather than the government identified the market and developed the land, had a great appeal and was adopted by the Dutch with a more market-oriented approach.

Learning From Mistakes and Successes

From the early 20th century until today in the Netherlands, the major generative role played by the government in the process of planning, designing, and constructing housing and settlements created a consciousness of the national successes and failures on the part of the professionals, the government agencies involved in the process, and the public. This was reinforced by continual planning for the next iteration, and generated an ability to make corrections in the process as it proceeded. Furthermore, because public money was spent, the press assiduously addressed new developments, thereby creating a national discussion.

Thus, in the last 40 years the cities that were deteriorating received investment in the 1980s, and policies for developing in suburban areas were altered in the 1990s to respond to the Fourth Government Report on Spatial Planning (Vierde Nota over de Tuimtelijke Orendeing Extra).[17] The initials of this report, VINEX, have come to stand for the areas created in 1990–2005 in light of the new policies.[18] Thus, too, projects once deemed failures—like Bijlmermeer and Lelystad—received corrective attention. New approaches to housing and community design—like the East and West Docklands of Amsterdam, the Kop van Zuid area of Rotterdam, or the application of the New Traditionalism in De Resident in The Hague—became issues of national attention.

It is not to say that in the United States we are unaware of our failures and do not experiment with new things. We too have formerly failing projects that have been radically changed, like Pruitt-Igoe in St. Louis, Cabrini-Green in Chicago, or Yerba Buena in San Francisco, and relatively new experiments in New Urbanism such as Seaside and Celebration. But these projects remain largely the domain of the local government, the professional designer, and the developer, receiving sporadic attention by the press and almost none by the national government. Only in the case of disasters like Hurricane Katrina does the national government or national press get involved, and then without the ability to implement solutions in long-term planning policy.

The CIAM-style Bijlmermeer buildings. Photo by Joseph Messier

A 1990s rowhouse development in Bijlmermeer

The role played in housing by the US national government has been primarily financial, providing the wherewithal for middle-class Americans to purchase a house. While in the early to late 20th century (1920s to 1980s) the national government played a small but important role in creating and supporting housing policy for low-income Americans, the primary focus was on the quality of the house itself, with only occasional awareness of the potential for addressing the housing context (such as the Model Cities project of the 1960s and early 1970s, which used housing to revitalize areas of urban decay). Today at the federal level in the United States, it is assumed that as long as financial support is provided for individual houses, there is no need for national concern that we live in well-designed communities that support such national goals as public transportation, clean air and water, reduction in pollution, and energy efficiency. National planning is haphazard at best. Each state and municipality is on its own, isolated from the next, and, for the most part, without mechanisms for coordination across boundaries, and for sharing what works and what is of dubious value.

Serving as correctives to isolation in the United States are organizations such as the Mayor's Institute and the Urban Land Institute, which provide an opportunity for local government officials and local planners and developers from across the nation to meet and share knowledge and expertise. But without a national policy or public support, these efforts remain largely invisible to the public and have limited effectiveness.

An essentially Dutch approach to housing and urbanism has been the willingness to experiment and to learn from the experience. It remains a question whether this will continue, as it seems very likely that the willingness was allied with the government's role in construction, thus reducing economic risk for individuals. Unless some solution is found to reduce the economic risk of exploration, the new market orientation of this sector may result in the loss of the Dutch innovation.

Abundance and Scarcity

A significant difference between the Dutch and American cultures can be understood by considering the approach to land development. With the Netherlands one of the densest countries in the world, on a very small landmass, land is viewed as a scarce resource to be conserved. Every inch of fertile farmland is valued, as is every inch of woods, bog, and heath. In the United States, on the other hand, land is viewed as an abundant resource where its transformation from open space to housing is simply taken for granted as "progress" in action.

However, increasingly, there is a developing understanding in the United States of a need to be more resource-conserving in our approach to the design of our environments. As our roads increasingly bear larger amounts of traffic and commuter time is greatly increased, we may realize the advantage of public transportation, bicycle, and pedestrian paths to offer alternatives to using the automobile. As good farmland is developed for housing and industry equally with land

Groningen landscape showing the long thin traditional fields, ditches, and canals © Karel Tomei

The planning of new development in Leidsche Rijn in Utrecht incorporates a part of the green heart and a traditional village. Photo by Jarrett Anderson

of no use for farming, we may see the advantage of long-term land planning. As the cost of infrastructure increases, we may realize the value in building more densely to make such investments more efficient. All of these and many more sustainable practices have been part of the Dutch approach to planning and design that might be more acceptable to the United States if we as a culture realize the significance of seeing the world as having limited resources.

Dutch and American Housing Projects

There are a variety of reasons that Dutch housing is different from housing normally built in the United States. For example, housing regulations are based on different assumptions. More windows and other openings are required to meet the Dutch standard for light and air. Since deep units do not provide enough access to the exterior, building blocks tend to have a much narrower footprint. Additionally, fire regulations in the Netherlands are based on preventing smoke inhalation, rather than preventing combustion as in the United States. Thus, whereas the Netherlands favors corridors with access to the outdoors between exit stairs, the United States favors single-loaded corridors that can be closed to prevent access to oxygen. Therefore, in the Netherlands single-loaded exterior corridors (or galleries) are commonly found, and access to the outdoors before reaching fire stairs is required in double-loaded corridors. Other less straightforward differences include unit size and a willingness to mix income levels in a complex. Perhaps due to land and space scarcity, typically Dutch units are 20% smaller than comparable US units.[19] While still experimental, as mentioned earlier, the willingness to combine different income levels is likely due to the lack of stigma attached to social housing, since people can remain in social housing after their incomes rise, paying market rate rents.

Additionally, Dutch attitudes toward wealth tend to minimize differences. Likely originating in Calvinistic traditions, the government has used tax policy to maintain a relatively small income gap between the richest and poorest citizens, and many affluent people seem to be uncomfortable displaying their prosperity. Even 17th–19th century manor houses built in the Netherlands are decidedly smaller than those built in other nations.

Finally, housing corporations have generated an orientation to creating long-term value in housing. The Housing Act of 1901 removed low-income housing from the speculative market, giving housing corporations a government mandate to implement housing policy.[20] The goal was to emancipate the working classes by maintaining affordable rental housing. Housing corporations invested in well-constructed buildings that they would subsequently maintain. In the United States, by comparison, often housing developers are not long-term investors who maintain a property for its lifetime, and thus may tend to use short-term investment strategies that result in lower-quality construction.

Since the 2008 recession, the situation in the Netherlands has changed. Although from 1900 through the VINEX period and up to 2005 the government created and implemented spatial plans every 5–7 years, no spatial plan has been created since 2005. The 2005 spatial plan represented a government move to a market-based system, and the recent national governments have not overseen its implementation; rather they have largely dismantled the national planning agencies. This may mean that the values represented in the housing discussed here may no longer apply to housing built more recently and to be built in the future.

Dutch and American Housing Units

While housing unit arrangement in dense housing in the Netherlands, in Europe in general, and in the United States are more similar than different, they are based on different cultural approaches. If we take mid-20th-century housing design as a baseline (because after World War II a great deal of housing was built and plans became more or less standardized), in all Western countries the dwelling unit is divided into independent rooms. The differences emerge in what these rooms are and how they are located.

Across the Western countries it is common to have three types of rooms (in this context corridors are not considered rooms but circulation spaces): bedrooms where people sleep; living rooms and dining rooms, where people gather (by mid-century combined into one space); and spaces linked to plumbing such as kitchen, toilet, bathroom, and/or laundry. In two-story dwellings in the West, the bedrooms are typically separated from and located on the level above the downstairs gathering spaces and kitchen. The differences within the countries arise in the arrangement and location of the hygiene spaces and the relations between the gathering spaces and kitchen, especially when the unit is a flat, all on one level. In all Western countries there is an increasing tendency for the living room, dining area, and kitchen to merge into a single space that may or may not be articulated to create implied sub-areas.

In the Netherlands one primary hygiene space is the WC, which contains the toilet and a sink. If there is only one toilet in the house, it will be located near the entrance. A second primary hygiene space for bathing and washing, the bathroom, may include a laundry space, and does not usually include a toilet. In a two-story residence this bathing/laundry space will often be upstairs, with the WC downstairs. This contrasts with the US "bathroom" that typically includes a toilet, sink, and shower or tub. If there is only one bathroom in a two-story US dwelling, typically a house, it will be located upstairs with the bedrooms. If a toilet is located downstairs it is called a half bath, even though it does not involve any bathing at all but consists of a toilet and sink, identical to what is called a WC in Europe. The explanation my Dutch friends gave me for the ubiquitous location of the WC next to the front door is that any workman coming to the house will be able to use it without going into the intimate bedroom area. They explain the separation of toilet and bathing as being a more sanitary approach. Another explanation relates to the sequence of introducing plumbing in the unit. When plumbing was first introduced in urban areas in the Netherlands, bathing took place in communal baths, and the WC was located in the corridor so it could be shared by several units. When each unit was provided its own toilet, it stayed in the corridor near the entry, and when the bath was added later as its own space, it was placed in the intimate area of the unit near the bedrooms, which in a house was typically upstairs.

Perhaps the biggest difference between contemporary housing designs in the Netherlands and the United States is found in the arrangement of flats (see Dutch

A typical contemporary Dutch flat

A typical contemporary American flat

and American unit plans and syntax gamma diagrams).[21] In the Netherlands, a corridor, what Ebner calls a "living hall" ("a central room to which all others are docked") orders the unit, as seen clearly in the diagram.[22] One enters into the corridor from which all the spaces radiate: WC/toilet, kitchen, coat closet, utilities room, storage area, kitchen-living-dining room (designated with a dark circle), bedrooms, and the bathroom (interestingly, unlike the United States where the doors open into the rooms, here many of the doors open into the corridor so it needs to be quite wide). All of the spaces have a similar relation to the corridor, thus the rooms are seen as flexible in relation to each other and dwellings are rented or sold in relation to the number of "rooms," in which is included the living-dining space, the bedrooms, and any other large non-kitchen or hygiene space in the unit.

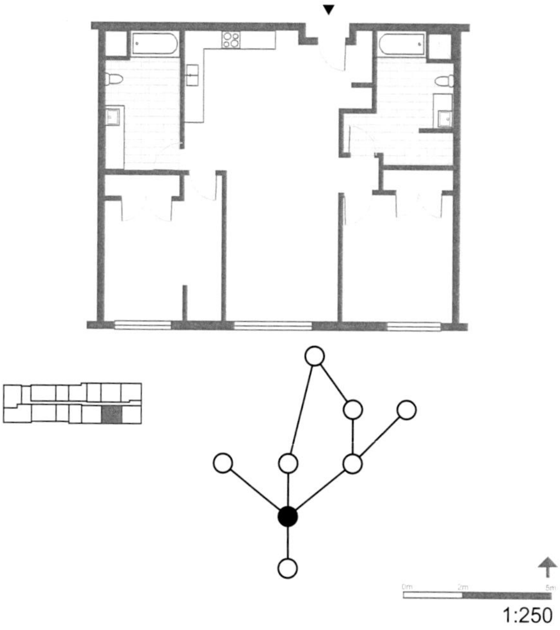

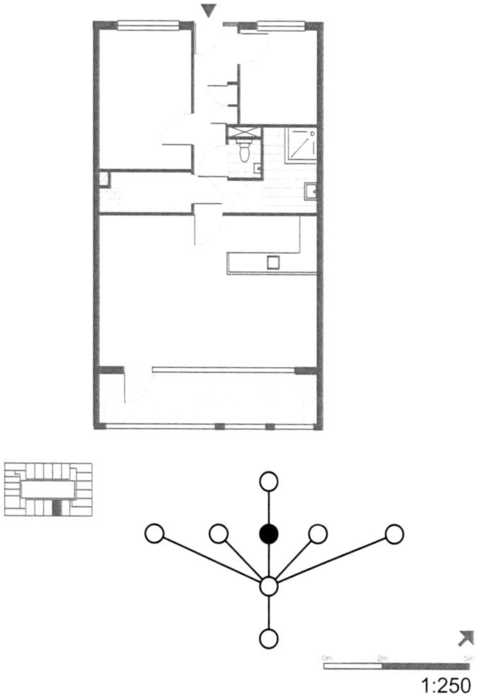

In the US flat, although some are organized with a living hall, just as often the entry area is not a separate space, but as shown in the illustration is integrated with kitchen-living-dining space, and links to a coat closet and half bath (if there is one). There is usually another open corridor leading to bedrooms and bathrooms, often accessed through the living room as with this example. There may be two bathrooms in the American flats, in which case one is very likely accessed through the corridor and the other is reached indirectly through a master bedroom. It is assumed that all dwellings will have a kitchen and living-dining spaces, so in the United States one rents or buys a house based on the number of bedrooms. By contrast, the Dutch system of designating rooms uses habitable spaces (including living rooms, for example) to designate rooms, so what is a three-bedroom apartment in US terms will be a four-room apartment in Dutch terms. In this book we will use the American system for discussing the number of rooms in a unit.

For dwellings with more than one story, another important difference between the Dutch and American dwelling is the staircase design. Perhaps because the Netherlands is such a densely populated country, one finds very steep staircases that in some historic buildings resemble ladders more than stairs. Consequently the Dutch standards allow a steeper staircase than in the United States, which means they take up less space in a unit. Thus a Dutch design that includes stairs cannot be directly translated to an American application.

As mentioned earlier, a key consideration in the design of dwellings is responding to the local codes and regulations. In most Western countries "habitable rooms" must have access to light and ventilation, although the designated amount may vary. In the United States in recent years, since the redevelopment of industrial "loft" spaces to housing, in some localities interior rooms without access to windows have been allowed to be called "bedrooms."

Another interesting design difference between the Netherlands and the United States is the degree of completion of a unit. In the Netherlands a dwelling unit will typically be sold without an installed kitchen or bathroom and without built-in furniture in the living room or elsewhere. Bedrooms will not include closet space and bedroom furniture is assumed to include an armoire with closet space. In contrast, a key sales point for US residences is the character of the kitchen and bathroom built-in counters, cupboards, and appliances, and closet space for the bedroom. In many American locales the designation "bedroom" is only given to a space that includes a closet.

Complex Housing

From the perspective of urban living, including walkability, bikeability, public transportation, and public infrastructure, there are many reasons for encouraging the construction of dense housing. Nevertheless it is important to acknowledge that the housing typically associated with density may not be popular with citizens, especially in the United States. Dense housing is seen as creating such attendant problems as traffic congestion and shadows on adjacent areas, as well as being unattractive because it inclines to be large, monolithic, and uniform in appearance. Resistance is even greater if the housing is all low-income, as it is seen as potentially reducing property values.

A key reason for the interest in complex housing is that it achieves density within a different framework than typical dense housing. Complex housing is dense without large high-rise buildings, without uniform and monotonous façades; it includes building functions other than housing and a mix of incomes and lifestyles.

Complex housing is a distinct form of dense residential design.[23] It is related to hybrid buildings,[24] in that it typically encompasses non-housing functions (usually commercial and/or civic) while providing moderate or high housing density. In addition to the mixed-use characteristics typical of hybrid buildings, a complex housing project typically incorporates:

1. Units both for rental and for purchase
2. Units for low-, moderate-, and high-income residents
3. Three or more types of dwelling units (e.g. rowhouse, maisonette, live-work unit, group home, flat, penthouse)
4. Diverse organizational strategies (layers, stacking, various types of access, use of courtyards, etc.)
5. Significant urban intervention
6. Height predominantly 4–8 stories but sometimes including a higher section of no more than 12 stories
7. Outstanding architectural design.

While complex housing is found in the Netherlands (mostly in the densely populated Western Randstad area) and in other Northern European countries such as Denmark, it is rare elsewhere. This approach to dense

Statistics: eight Dutch complex housing projects (arranged by density)

	De Muzen	**Vrijburcht**	**De Zilvervloot**	**Carnisselande**
Project	De Muzen	Vrijburcht	De Zilvervloot	Carnisselande
Architect	Atelier Pro	CASA	AUAI	Steenhuis Bukman
Developer	Groen Stad Almere	De Key	Woonbron	Estrada Projecten
Date	1997	2007	2005	2006
Location	Almere	Amsterdam	Dordrecht	Barendrecht
# Units	167	52	130	171
Density	80 units/ ha	120 units/ha	170 units/ha	190 units/ ha
	30 unit/acre	50 unit/acre	70 units/acre	80 units/acre

housing contrasts with that typically taken in the United States and in many other countries, in which a project is formed by lining up flats (one-level apartments) along a double-loaded corridor and layering these multiple identical floors to create a building made up exclusively of one type of dwelling.[25]

Selecting the Eight Case Studies

The choice to study complex housing responded to a fascination with Dutch housing and particularly housing blocks that organized diverse housing types in an organic way rather than the repetition of identical elements typical in many countries, including the United States. Certain obvious differences, like the variety in the façades generated from the juxtaposition of different housing types, raised questions about the way that the Dutch architects created the designs and how the developers in the Netherlands were able to build what seemed, to an American eye, such "inefficient" designs.

Initial analysis of the first collection of housing projects led to the preceding definition of complex housing, and the realization that not all of the projects met all of the criteria, and in fact several projects met only a few. Because categories are best defined by central characteristics rather than by boundaries,[26] and few of the cases met every single criterion, for this study each case was selected to meet most but not all of the criteria. Application of the criteria ruled out several projects, with six originally chosen for detailed analysis. Once research began and the search for cases expanded, the number of cases was increased to eight for sufficient variety. The final selection includes urban and suburban examples, smaller and larger projects, a range of density from moderate to very high, a range of non-housing functions, examples that were designed with various types of participatory process and without, and examples with a range of formal approaches including one traditional design.

Contents of the Book

Complex housing represents a particularly Dutch solution to the design of dense housing in its orientation to the urban block and organic mix of design elements. To help

Statistics: eight Dutch complex housing projects (arranged by density)

	De Beeklaan	**De Opgang**	**La Grande Cour**	**Silodam**
Project	De Beeklaan	De Opgang	La Grande Cour	Silodam
Architect	SCALA	KCAP	MVSA	MVRDV
Developer	Bouwfonds	NL Development	Bouwfonds	De Key
Date	2007	2010	2008	2003
Location	The Hague	Amsterdam	Amsterdam	Amsterdam
# Units	105	79	253	157
Density	200 units/ha	240 units/ha	350 units/ ha	600 units/ ha
	80 units/acre	100 units/acre	140 units/acre	245 unit/acre

the reader appreciate this approach, this introduction provides background on the history, culture, and politics of the Netherlands. The subsequent chapters provide an analysis of the housing to reveal its design principles and, finally, the basis for a discussion of its replicability in the Netherlands, the United States, and other countries.

Chapter 1, "Analyzing Complex Housing: Typology," addresses the research methodology, explaining the use of the case study approach, typology, space syntax gamma analysis, and density. It frames the typological analysis that generates the design principles presented in Chapter 10.

Chapter 2 is about De Muzen, a suburban project built by Atelier Pro Architekten in Almere in 1997 as the focal point of a large park. The earliest of all the projects, this very long linear building was originally designed to serve older adults, but the for-purchase units ended up being sold on the open market. It comprises 167 units with a density of 77 units per hectare (31 units per acre), and parking on grade.[27] This project shows how a very long landmark building, originally designed as housing for seniors, can fit within a suburban neighborhood at the same time that it provides an interior courtyard and a variety of dwelling types.

Vrijburcht, designed by CASA in 2007, is the subject of Chapter 3. Designed using participatory and green

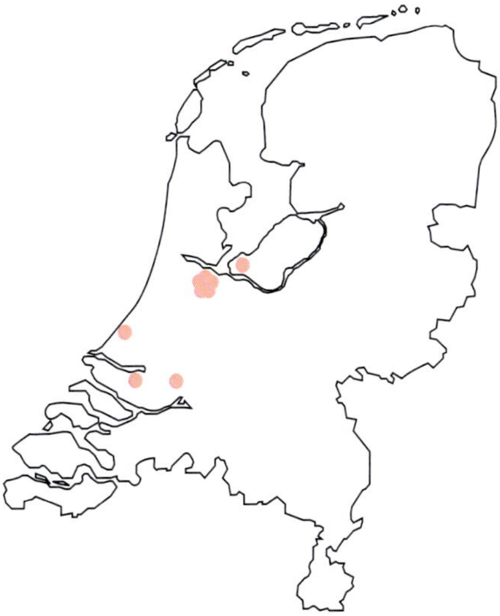

Locations of the eight projects in the Netherlands

design, the project includes 52 units at a moderate density of 120 units per hectare (50 units per acre). Its central courtyard is the social heart of the project. Within the complex are a group home, a childcare center, a theater, a café, and a community center. Vrijburcht demonstrates that participatory design and concerns for developing community assets can generate a project with a vibrant residential collective in a courtyard setting that serves the larger neighborhood.

Chapter 4 concerns De Zilvervloot, a 125-unit project in Dordrecht with a density of 170 units per hectare (70 units per acre), designed by Atelier d'Urbanism, d'Architecture et d'Informatique in 2007 in a participatory process that renovated a neighborhood in Dordrecht built in the 1950s. Designed as a set of nine buildings that surrounds two interior courtyards, the ground level is devoted to commercial activities with parking below. Among the many lessons from De Zilvervloot is the manifestation of the potential for a well-planned, participatory urban project with greatly diverse unit types to reinvigorate a declining neighborhood.

Chapter 5 examines Carnisselande, a 2006 project in suburban Barendrecht. Designed by Steenhuis Bukman Architects, the complex of 190 units per hectare (80 units per acre) marks the town square with an eighteen-story housing clock tower. In addition to social housing, moderate-income housing and luxury housing, it incorporates a library and a clinic and commercial spaces on the ground level, as well as on-site parking. Carnisselande encompasses many types of housing including a group home, and has become an urban landmark for its suburban community.

The subject of Chapter 6 is De Beeklaan, a traditional design built in The Hague by SCALA Architecten in 2007 at a density of 200 units per hectare (80 units per acre). The 105-dwelling project, complete with a landmark tower, is designed to fit its 19th-century neighborhood context. It incorporates commercial space, interior and exterior at-grade parking, and an assisted living project for seniors as well as flats and rowhouses for families. De Beeklaan demonstrates how a traditional design can serve contemporary housing needs at the same time it restores a neighborhood.

De Opgang, described in Chapter 7, is a project located in Amsterdam and designed in 2010 by KCAP Architects&Planners, with 79 units at a density of 240 units per hectare (100 units per acre). Developed by a church group in conjunction with a private developer, it does not include social housing, but has low, moderate and upper-income apartments as well as townhouses. In addition to housing, this project incorporates underground parking, the church and a childcare center as well as a three-level interior courtyard. De Opgang shows how the cooperation of a private developer, a government official, and a church group in the development of housing can maintain a religious organization as an essential part of an urban community.

Chapter 8 analyzes La Grande Cour, a project designed by three architectural firms—MVSA Architects, de Architekten Cie and Heren 6 Architecten—and built in 2008 in the West Docklands of Amsterdam near the railroad station and on the IJ River. This project, which has the most units at 253, and a density of 350 units per hectare (140 units per acre), is designed around three courtyards. The ground level is devoted to commercial activity with parking underground and housing above. This project is an example of the Dutch designers' and developers' ability to create livable urban housing at a high density for all income levels through careful articulation of the building massing.

The focus of Chapter 9, Silodam, is located in the West Docklands of Amsterdam, on the IJ River. Designed by MVRDV Architects in 2003, this project has 157 units at a density of 600 units per hectare (245 units per acre). The structure has four sections in a ten-story layer cake of different types of units with commercial space at the base, a public terrace overlooking the river, and adjacent automated underground parking. An exception to the complex housing height criteria, this project is a tour de force in terms of locating many different housing and unit types in one dramatic, very dense building, and providing all residents with light, air, and river views.

Chapter 10 presents the design principles that are used to create complex housing in the Netherlands. It discusses the patterns of similarity within the group of housing as well as the differences that make each project unique.

Finally, Chapter 11 returns to the original question of what can be learned from studying the eight Dutch examples. Based on lessons taken from the projects, and views of housing professionals interviewed for the book, it addresses how the circumstances in the Netherlands that resulted in complex housing relate to planning, development, and financing as well as to urban and housing design elsewhere in the world. It discusses such issues as the role of planning in creating the context for design and the implications for creating complex housing when there is little or no planning; developers being motivated to create such unique buildings because they become icons for their companies; and how long-term ownership of buildings relates to high-quality design.

Notes

1. Richard Stolzenburg, Class Lectures (1999–2004), based in part on the book *Stad en samenleving* by Kees Doevendans and Richard Stolzenburg (Groningen: Martinus Nijhoff, 2000).
2. Centraal Bureau voor de Statistiek (CBS). http://statline.cbs.nl/Statweb/publication/?VW=T&DM=SLNL&PA=71527NED&D1=5-7&D2=a&HD=151230-1204&HDR=T&STB=G1; ttp://statline.cbs.nl/Statweb/publication/?VW=T&DM=SLNL&PA=81955NED&D1=0–2,7&D2=1&D3=0&D4=16,33,50&HD=151230–1206&HDR=G1,T&STB=G2,G3. I say "over 7.5 million housing units" because data have been collected only since 1921.
3. Ministry of Housing, Spatial Planning and the Environment. 2004 "Summary national spatial strategy: Creating space for development," *Ministerie van Infrastructuur en Milieu website (Bestelcode 4054)*. (www.rijksoverheid.nl/documenten/brochures/2010/11/23/summary-national-spatial-strategy-creating-space-for-development
4. Sections of this chapter that have an asterisk include material presented in Richard Stolzenburg Class Lectures (1999–2004).
5. Wouter Reh, Clemens Steenbergen, and Diederik Aten, *Sea of land: The polders as an atlas of Dutch landscape architecture* (Wormer: Stichting Uitgeverij Noord-Holland, 2007).
6. Donald Grinberg, *Housing in the Netherlands: 1900–1940* (Culemborg: Delft University Press, 1982), 20.
7. Ibid., 34.
8. Nancy Stieber, *Design and society in Amsterdam reconfiguring urban order and identity, 1900–1920* (Chicago: University of Chicago Press, 1998), 73.
9. Hendrick Ploeger, Aart ven Velten, and Jaap Zevenbergen, *Real property law and procedure in the European Union: Report for the Netherlands* (Delft: OTB Research Institute, 2005), 7.
10. Grinberg, *Housing in the Netherlands*, 42.
11. For a history of the Dutch block see Susanne Komossa, Han Meyer, Max Risselada, Sabien Thomaes, and Nynke Jutten, *Atlas of the Dutch urban block* (Bussum: THOTH, 2005); and for an understanding of how it has functioned see Susanne Komossa, *The Dutch urban block and the public realm: Moels, rules, ideals* (Nijmegen: Vantilt, 2010).
12. Komossa et al., *Atlas of the Dutch urban block*, 2005.
13. Vaughn Aubuchon, "Vaughn's Countries of Europe Summary," Vaughn's summaries, www.vaughns-1-pagers.com/geography/european-countries.htm.
14. Drawing after Wikipedia map "Population of the Netherlands 2006" based on information from the Centraal

Bureau voor de Statistiek (Demographics of the Netherlands), https://en.wikipedia.org/wiki/Demographics_of_the_Netherlands 2016.

15. Aedes Vereniging van Woningcorporaties, *Dutch social housing in a nutshell* (Hilversum: Aedes, 2003), 6.
16. Ibid., 7.
17. The role of the policy in the 1990s is described by Maaike Galle and Ettjen Modderman, "National spatial planning policy in the Netherlands during the nineties," *Netherlands Journal of Housing and the Built Environment* 12, no. 1 (1997), 9–35.
18. Jelte Boeijenga and Jeroen Mensink, *Vinex Atlas* (Rotterdam: 010 Publishers, 2008), 3.
19. Julia Robinson, *Student research projects comparing Dutch and US housing projects, architecture 5750: Contemporary Dutch housing in urbanism* (unpublished papers, Minneapolis: University of Minnesota, 2007–2010).
20. Grinberg, *Housing in the Netherlands*, 38.
21. Julia Robinson and Mandana Motamed, "Comparative analysis of housing from the Netherlands and the United States," paper presented at the International Association of People-environment Studies (IAPS), (Lund, Sweden, June 28, 2016).
22. Peter Ebner, *Typology +: Innovative residential architecture* (Basel: Birkhäuser, 2010), 114.
23. Robinson, Julia W. "Dutch complex housing: Design for density," in *Future of architectural research: Architectural Research Centers Consortium, 2015,* edited by Ajila Aksamija, John Haymaker, and Abbas Aminmansour (Perkins+Will, 2015), 412–419.
24. Javier Mozas, "This is hybrid," in *This is hybrid: An analysis of mixed-use buildings*, edited by Aurora Fernandez Per, Javier Mozas and Javier Arpa (Vitoria-Gasteiz: a+t, 2011), 12–25.
25. Robinson, *Dutch complex housing, 2015*, 421.
26. Amos Rapoport, "Defining vernacular design," in *Vernacular architecture: Paradigms of environmental response*, edited by Mete Turan (Brookfield: Avebury, 1990), 68–102.
27. The calculation method for determining density in this text is explained in Chapter 1.

1 Analyzing Complex Housing: Typology

VRIJBURCHT

This chapter frames the investigation of complex housing by addressing key concepts and approaches such as case study method, space syntax, gamma analysis, density, and typology. To reveal the design principles that generate Dutch complex housing, eight examples were chosen to analyze in depth, with the goal of understanding the commonalities of planning, design construction, and maintenance of dense housing in the Netherlands. Focusing on eight cases rather than addressing Dutch housing and urbanism broadly, the study examines the projects as actually designed and implemented. The specifics of each example reveal different, important issues that can be best understood in the light of a project's history and design decisions. To learn whether and how a project might be reproduced somewhere else, it is essential to know the factors that led to its inception and construction. Finding comparable projects with a range of approaches was essential for this effort.

General Terms and Usage

Case Study Approach

The case study methodology is a normative architectural research practice. In a field that proposes future environments, looking at past projects can be practical and inspiring. Learning what has worked, what has been unsuccessful, what are the new ideas, and so on yields potentially important directions for new projects. In his book on case study, Yin defines a case study methodology as an in-depth investigation of a contemporary phenomenon in a "real-life context" for which the boundaries are unclear between the phenomenon and its context. He further suggests that the complex variables inherent in such an approach require convergence between multiple sources of evidence to assure reliability and that this type of study "benefits from having a strong theoretical position to drive it."[1] Because many factors contribute to the successful implementation of a housing project in a "real-life" context, and because these factors can be identified and compared across projects, case study is an ideal method for this investigation, with typology providing the strong theoretical position.

Case studies in architecture focus on a great variety of issues, and there is a rich history of approaches that have been taken, perhaps beginning with Vitruvius (c. 22 BC), whose more anecdotal approach was replaced in the Middle Ages with careful analysis, such as that by Alberti (1485) and Palladio (1570) and supplemented in the 19th century with rigorous typological methodology.[2,3,4] Although it can be useful to look at only one case, comparative analysis, wherein several cases are analyzed in relation to each other, offers the possibility to see larger patterns and understand how projects are typical, innovative, or are outliers due to any number of reasons. Setting projects side by side that have similar goals may also reveal unexpected assumptions and issues that allow a greater understanding of all the cases under investigation.

Complex Housing Cases

As explained in the introduction, eight case studies of complex housing were chosen with reference to the definition, and the definitional approach taken was one of centrality rather than defining borders; cases had to meet most but not all of the criteria. Indeed, few of the cases meet all of the criteria. They represent a mix of architectural styles and organizational strategies, including both urban and suburban locations.

The eight examples chosen for study are all within the prescribed density (over 70 units per hectare (28 units per acre)), built between 1990 and 2010, and are in the Randstad area where there is an interest in creating higher density. They represent a variety of approaches and factors including a mix of size (52 to 253 units). All of the projects have units both for purchase and rental (although the rental units at De Vrijburcht are in the

Table 1.1 Dutch complex housing selection criteria

Selection Criteria								
Criteria	**Other functions**	**Density: Units/ Hectare (Units/acre)**	**Rental & Purchase**	**Income Levels Served**	**# Hsg Types**	**# Unit Types**	**# Access Types**	**# units**
Project								
De Muzen	YES	77 (31)	YES	2/3	2	8	3	167
Vrijburcht	YES	118 (48)	YES	3/3	5	5	3	52
De Zilvervloot	YES	170 (69)	YES	3/3	5	10+	4	130
Carnisselande	YES	193 (78)	YES	3/3	4	9	3	171
De Beeklaan	YES	197 (80)	YES	3/3	2	7	2	105
De Opgang	YES	242 (98)	YES	2/3	2	8	2	79
La Grande Cour	YES	351 (142)	YES	3/3	3	10+	4	253
Silodam	YES	604 (244)	YES	3/3	4	8	4	157
AVERAGE		244 (99)			3.5	8+	3	139

group home). Six of the projects incorporate social or low-income housing as well as moderate- and upper-income housing (the exceptions being De Opgang and De Muzen, which do not have luxury units). One project includes affordable housing and rental units but not social housing for rent (De Opgang). Seven projects have commercial, civic, or ecclesiastical functions (De Muzen has a community space that serves its own housing community). Finally, all of the projects are compelling because of their development history, for example De Opgang, the least dense project and the site of a church, is the only project with a private developer.

Density

Because dense communities are more efficiently served by infrastructure and transportation, the worldwide concern to reduce the carbon footprint has generated interest in increasing the density of our urban areas, especially in relation to public transport, where transportation experts recommend an overall urban density of around 74 units per hectare (30 units per acre) for the support of light rail, and about 111 units per hectare (45 units per acre) for heavy rail.[5]

The term *density* is the subject of much discussion, as addressed in Pont and Haupt, who present various ways to calculate density and address in detail the advantages and limitations of each method.[6] It is important to observe that a given numerical density for housing can be implemented in a variety of ways, depending upon the type of housing chosen (high-rise, rowhouse, walk-up apartment), especially at low to moderate densities.

The method of calculating density used in this study is similar to what Pont and Haupt and others call *net housing density* or a calculation that "excludes all public streets,"[7] that is the number of dwelling units per area at the housing site. The calculations we used were based on the site information we received from the architects. We also calculated density using the housing footprint, calculated from the architects' plans rather than size of site, although those figures are not presented in this text.[8] While these calculations do not take into account the actual number of people living in a given area or the size of the unit (since a large unit and a small unit are counted equally), they are relatively easy numbers to calculate and to understand, therefore net housing density is commonly used by designers. Net housing density rarely reflects the actual density of a given area of the city, because neighborhoods do not just include housing, but may include extra open spaces, extra wide streets, parking areas, and so forth. Nevertheless, housing designers are typically asked to create a certain number of housing units on a given site, and net housing density provides a way for a designer to compare different design alternatives in terms of units per hectare or units per acre.

For this project the number of housing units and the site area are calculated using information provided by the architects and developers, verified by checking the floor plans provided by the architects and site plans online. Some information provided by the architects and developers uses "units" to include not only the housing, but commercial units as well. Here, for consistency across the examples, we do not include the non-housing spaces as units, even though they do add to the overall site density in terms of site coverage.

In the past, density has been a rather controversial topic. Mitchell documents research in the 1950s and 1960s such as that by Loring that suggests that very dense housing conditions are related to social disorganization; and work by Gruenberg that indicates a relationship between areas with high density of housing and high admission rates for psychoses.[9] But Mitchell concludes that such effects are not directly related to housing itself, but have more to do with the neighborhood context of the housing. In the 1970s, as a result of the apparent failures of many high-rise housing projects for low-income people around the world, researchers identified density and crowding as a possible cause, leading researchers such as Stokols to make a distinction between crowding (a psychological response to density) and density itself (a numerically determined measure) that allows us to understand that a single density may elicit varying degrees of crowding in different people.[10] This idea was further developed by Amos Rapoport to differentiate perceived density from actual density, since different forms of housing that have a similar density (e.g. rowhousing and high-rise housing) may be seen as being more or less dense than their measurements would suggest.).[11] In this project, although we are not addressing it directly, perceived density is an underlying theme, and it is an untested assumption that Dutch complex housing will be perceived to be less dense than it actually is.

As discussed in Berghauser-Pont (2010), Jane Jacobs suggested there might be "thresholds" of density.[12,13] "Densities are too low, or too high, when they frustrate city diversity instead of abetting it," both in the sense of diversity of activities and diversity of housing.[14] Citing particular examples of successful urbanism, Jacobs advocates a minimum of about 125 units per acre and a maximum of 200 (300–500 units per hectare). According to Jacobs, densities above 200 are likely to require standardization of design, thus to have a more uniform urban pattern.[15] In contrast to Jacobs, more recent European authors such as Mozas and Per suggest lower thresholds, with low density as 50–100 units/hectare (20–40 units per acre), medium density as 100–200 units/hectare (40–80 units per acre), high density as 200–300 units/hectare (80–120 units per acre), and very high density as over 300 units per hectare (120 units per acre).[16]

Using the Mozas and Per definitions, of the projects discussed here—but recognizing that our method of density calculation using building footprint may tend to make higher densities than the method of calculation used by Mozas and Per—two examples are low density (De Muzen and Vrijburcht), two examples are medium density (Carnisselande and De Beeklaan), two examples are high density (De Zilvervloot and De Opgang) and two are very high density (La Grande Cour and De Silodam).[17]

Space Syntax Gamma Analysis

When studying buildings with complex arrangements, examination of the plans can be important but sometimes difficult to understand. Among the variety

Space syntax gamma diagrams of all eight projects

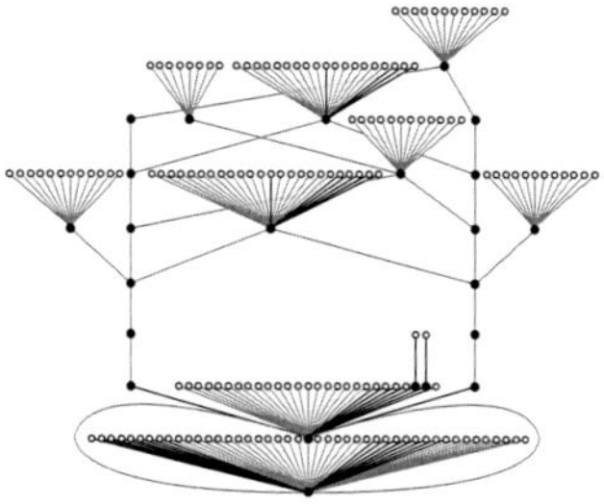

De Muzen

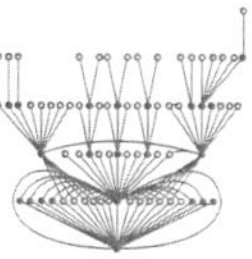

Vrijburcht

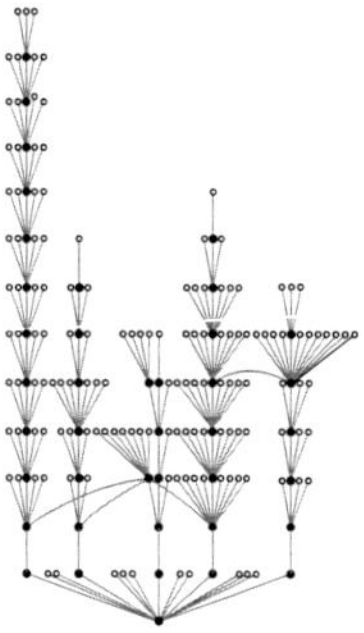

De Zilvervloot

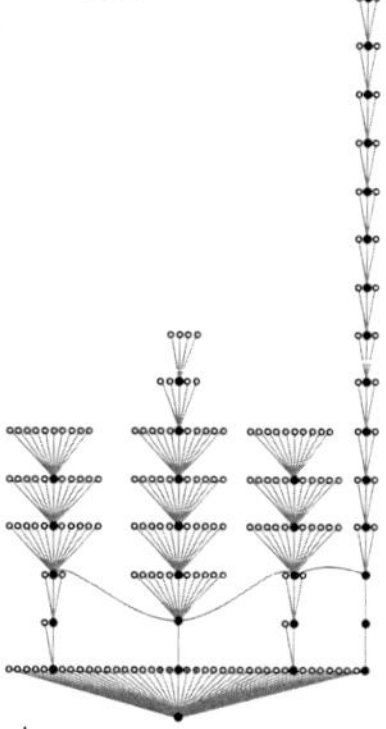

Carnisselande

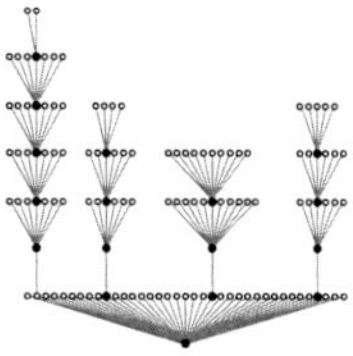

De Beeklaan

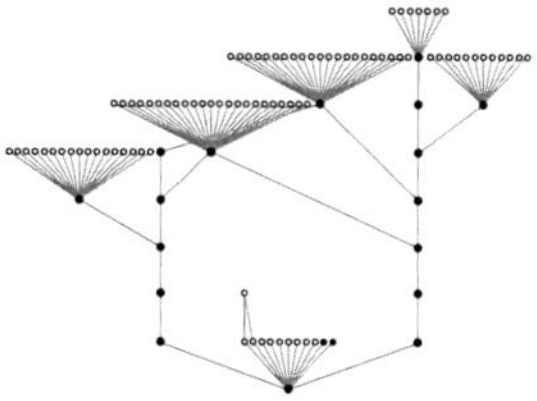

De Opgang

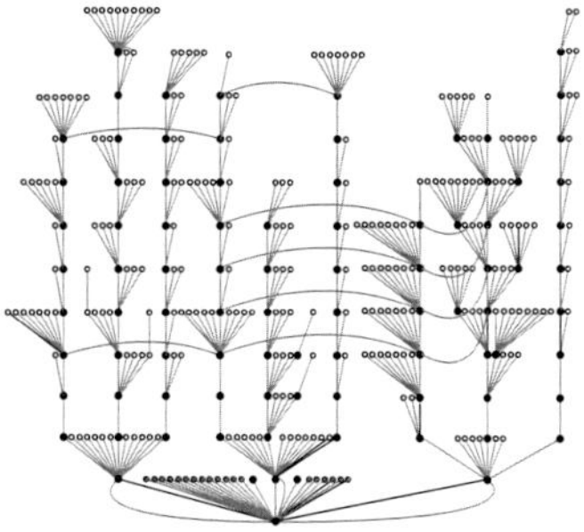

La Grand Cour

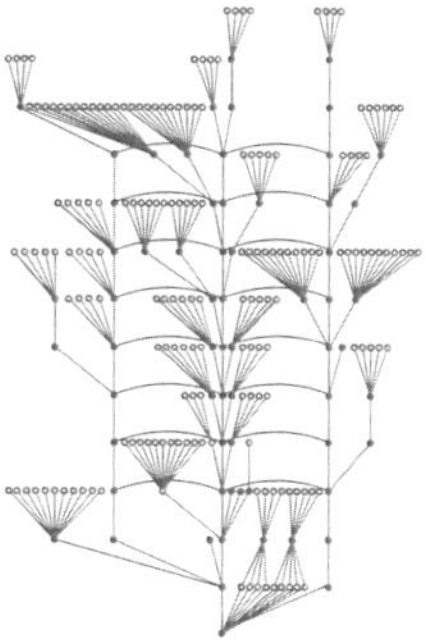

Silodam

Plans of all eight projects

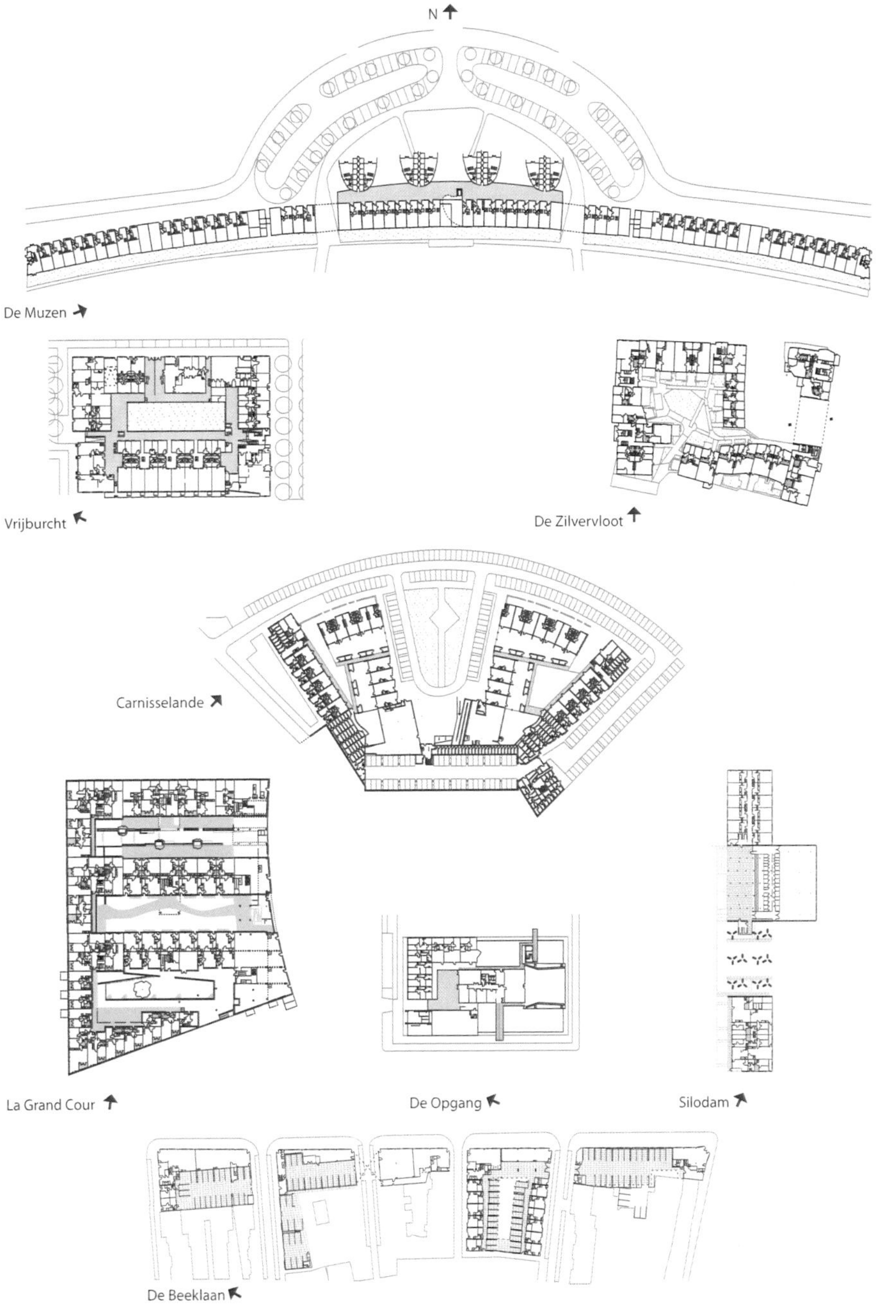

of techniques that have been used to abstract elements of plans, space syntax gamma analysis stands out as a powerful tool for studying the way spaces are structured. By focusing attention on the links between spaces rather than space size and shape or the character of the separating elements or openings, gamma analysis provides a distinctive view of the organizational structure of a building.[18]

Space syntax is a broad category of diagram-based analysis that derives from linguistics as applied to the study of urban patterns and architecture.[19,20,21] Since 1976, it has been formalized and developed into a variety of analytical tools, and applied in both building and urban analysis by Bill Hillier, Julienne Hanson, and others.[22] Of the many types of space syntax diagrams (e.g. axial maps, interface maps), in this study we employ gamma analysis diagrams—also called convex space analysis—to reveal important attributes of the various spatial arrangements.

There are no fixed conventions for creating gamma analysis diagrams. Instead, researchers define their mode of determining convex spaces and develop the diagrams consistently within their study to show what is salient about the environments they are analyzing, choosing for example the point of view (perspective of the visitor, resident, worker, etc.), definition of the boundaries of spaces (how rooms or corridors are defined), and how to represent a change in level.[23,24] In this study, for consistency we uniformly take a visitor's perspective toward the housing projects and begin (or justify) our diagram from the visitor's entrance. To focus on the organization of the complexes, the diagrams are limited to describing the public or semi-public circulation spaces, excluding dwelling interiors. The diagram reflects the visitors' potential journey to every dwelling, beginning from the street where they typically enter the complex; in most cases there are multiple entries. Each space is represented by one dot. The links between spaces are represented as lines. Our diagrams begin at the bottom and as each space is linked to another, create a new layer in the diagram.

The diagrams start from the exterior, shown as the lowest small circle in the diagram. Each space that is entered from the street is diagrammed as a circle on the line above, represented as the next layer. Each time a new space is linked to an old one, it is drawn on a successively higher layer in the drawing, with its connections to other spaces shown as lines. An entry space that connects only to a corridor space is shown as a circle on a lower layer connected to the corridor space on the next layer up by a single line. For example, if the corridor connects to seven spaces, its circle will have lines that connect the circle representing the corridor to seven circles on the next higher layer, representing the spaces that are entered from the corridor. While corridors that connect to each other through closed doors are shown as separate spaces, in this study even very long circuitous corridors are shown as a single space (typically these are exterior galleries and thus visible for the entire route).

Because the elevator is the primary vertical circulation in most of these buildings, the visitor experience is represented as if the visitor enters and exits the elevator on each successive level. Thus each level is shown as deeper within the building than the one below it. Where there are two elevators that connect to a single level, those elevators are both shown on the diagram as connecting between floors (for example, see the space syntax diagrams of De Opgang and Silodam).

When examining gamma diagrams, there are several things to consider: the total number of connecting spaces, the organization of the spaces on the site, and the complexity of the overall structure. The number of connected circles or dots suggests the scale of the complex overall (these diagrams exclude the spaces within the dwelling units). However, floor plans are a much better measure of spatial size than syntax diagrams. The organization on the site is indicated by the number of entrances to a given complex at the base of the diagram and the subsequent links to vertical threads, or circulation routes. In some cases, there are few entrances directly from the exterior, whereas in others there may be many. Some of these are entrances to individual dwelling units, others are entrances to the non-housing functions such as commercial or civic areas, and still others are entrances to all or part of the multi-family housing complex. Because the visitors do not typically use the interior parking for cars or bicycles, the parking areas are not shown in the diagrams. The complexity of the overall structure is indicated by the number and character of threads (spaces connected to a vertical circulation path) that exist side by side, or connected to each other. Within a given complex there may be few or many threads, and the threads may have a repetitive pattern or may be unique.

Typology

Deriving from linguistic analysis of language, typological analysis is based on the principle that thinking is fundamentally linked to the ability to categorize things and name them. In architecture, typology or classification by type breaks a design into constituent elements. The long history of typological analysis in architecture is usually seen as originating in the work of Durand, who in the 18th and 19th centuries studied buildings by classifying them in categories by use (e.g. temples, basilicas, forums) and elements (e.g. vestibules, columns) and setting drawings (primarily plans and elevations) of many cases of the same category or element side by side.[25] Although not all those who use typology as a tool for analysis would see it as a tool for design, Durand and others also saw the approach as a way to innovate by creating new elements and arranging known elements in new ways.

Typology was extensively employed to study architecture and urban design during the 1970s and 1980s, and it is being applied more recently to urban form and dense housing.[26,27,28,29,30,31] Since the 1980s space syntax has enhanced typological methodology with analytic diagrams and mathematical analysis of relationships between spaces in environments.[32,33]

A strong reason for selecting typological methodology in this project is that many Dutch architects consciously employ typology as a design strategy.[34] This project uses the method to study the design decisions that architects make about architectural form as defined by typological elements of the designs. These decisions are not independent of each other, but are based upon the designer and client criteria, and thus the values held for the project. Such decision-making creates the overall design approach of a particular designer or architectural firm. What Yin identifies as convergence of "multiple sources of evidence" is important as a strategy in multiple case studies.[35] Comparison of the patterns of design choices to interviews with the designers, to knowledge about codes and regulations, as well as to generally shared broad criteria about what makes good housing, reveals the design principles that architects in the Netherlands are following. The analysis shows which attributes the different housing complexes have in common and how they are each unique, thus distinguishing between design principles that characterize Dutch complex housing and design principles that create variety between exemplars.

Many physical design attributes could be studied, but in our preliminary work we found a subset that are particularly revealing. While we know that the investigation could be even more detailed, we have focused our attention on the following elements:

1. Location, context, and site
2. Building functions
3. Site design
4. Courtyards
5. Articulation of form

6. Housing type
7. Unit type
8. Access
9. Outdoor space
10. Syntactical structure.

It is important to acknowledge that typological elements are not pure. That is to say that they are not totally independent of each other, and in some cases may not be totally parallel in their place in an implicit typological structure. Because types are not defined by their edge conditions or boundaries, but by their central conditions, there are often ambiguities and lack of consistency within a type. So for instance, a maisonette is typically an apartment that has more than one level, and skip-stop apartments, because they have two levels or more, can be classified as maisonettes. However we would like to differentiate housing type not only by the internal arrangement of the dwelling, but also by the type of access. Therefore we distinguish between (plain) maisonettes that may be entered from a variety of access ways and may have various access to light and ventilation, and skip-stop units that are reached from a double-loaded corridor, and that have access to both sides of a building on a floor-through level above and/or below the corridor level. Thus the categories that we have selected are tailored to this set of examples.

The remainder of this chapter defines the typological categories being used. The purpose is to be clear about how the terminology is being applied so that the analysis is consistent. It is a challenge to create the categories because, as just seen, there are many ways things can be broken down into subordinate categories and grouped into superordinate categories. Also types are context driven, so that in different situations they may be categorized differently. In this case, housing types, as they are building types, are considered to be basic types (those objects or ideas that when encountered by people perhaps in an image are given a common name, like "car," "cat," "fork") and other categories are either parallel (also basic categories but of a different kind), superordinate, at a higher category level or scale, or subordinate, at a lower category or scale.[36] As a result, a given category might have a very different set of subcategories depending upon the context (house in terms of materials versus house in terms of arrangement of dwelling).

These types used here were created based upon literature about housing types,[37] interviews with architects and developers, and the analysis of the complex housing projects. This is not intended to be a definitive list of types, but simply a way to describe the housing found in this study. Although the types are intended to be generically descriptive, because they derive in large part from the observations of this study, their origins are to some degree tautological.

Typological Analysis—Location, Context, and Site

Context

Complex housing projects are large and are typically oriented in some visually prominent space, whether on a *wide street*, a *square* or other open urban space, a *park* or green space, or on *water*, whether a canal, river, or lake.

Project Functions

Every building in this study incorporates some form of non-housing function. A question that arises when putting diverse building functions into a single structure is the extent to which the functions require a strong expression of separateness or whether the overall structure should read more as a unified whole.

Housing

As a key goal of the VINEX approach was to bring together into one building people of different incomes and lifestyles, design considerations include sufficient security, privacy, and a sense of personal control that permit comfortable mixing between groups. One organizational issue is how much to limit the numbers of units accessible from a single entry, elevator, and/or corridor, and whether to make some units directly accessible from the street or from the courtyard. Additional considerations include how much to visually differentiate between housing types and units, and how to design the dwellings' appearance so that the difference in resident income is not conspicuous. For more discussion of housing and unit types, see "Typological Analysis—Housing Type and Unit Type" later in this chapter.

Commercial

Housing large numbers of people in a single block offers opportunities for commercial ventures within the building, both by the residents themselves in the form of live-work units, and also for services such as offices, restaurants, cafés, or supermarkets. Typically these functions are located at the ground level. Because commercial activities serve people living in the adjacent housing as well, they offer a way for a project to be integrated in its neighborhood.

Community or Civic Space

Similarly, community spaces such as a childcare center, church, theater, interior atrium, and library spaces found in these cases are usually located on the ground level and serve not just residents, but also the whole adjacent community. These functions also can make the building an essential part of its neighborhood.

Parking and Storage

In the Netherlands, projects include both automobile and bicycle parking as well as storage for residents. Because public transportation and bicycle use is so highly developed in the region being studied, one parking space per unit is deemed sufficient. Storage space outside the dwelling (often underground or on the ground level) is very important because units in the Netherlands are typically very compact.

Site Design

The projects in the study are on both suburban and urban sites. They vary in terms of size, height, massing, articulation of form, and courtyard design.

Size and Height

Building size can be described in terms of area and height. In the Netherlands, while high-rise buildings exist, they are found in very few places. The buildings being explored here are mid and low-rise buildings, but have a high density because they are efficiently laid out on the site. But they are chosen for examination because the efficiency is not based solely on repetition of similar elements. Even more interesting than size is how the building is configured. One way of considering building configuration is to describe a building's massing.

Massing

There are many ways to describe massing, but for the housing we are studying three common massing forms can be identified: the tower, the slab, and the perimeter block (of which variants surround all or part of an interior open space). These forms are not totally separable, however, as the perimeter block may be a pure form or may be a form composed of slabs and towers.

Types of massing

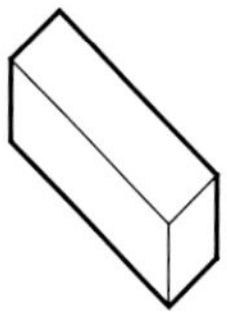

a. Slab

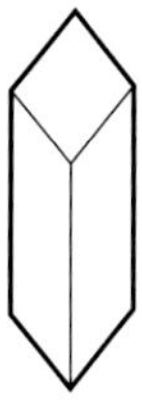

b. Tower

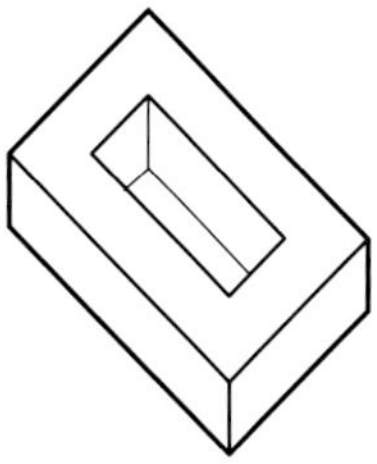

c. Perimeter Block

When housing is thought of as buildings, as typical in the United States, there is a tendency to associate the massing with the ground or site, so that the project is a single massed form from the ground up. In places where housing is thought of as an urban fabric, the massing may be disassociated with the ground plane, and perimeter blocks may be created from combinations of slabs and towers by layering, suspending and/or stacking slabs alongside towers.

Tower

Towers are formed by stacking housing units to create a tall, narrow structure. In the Netherlands, this type of massing typically has a limited number of units per floor (three or four) that are accessed from an internal hall or vestibule. Having a small number of units on each level, as is typical of the tower buildings found in this study, offers the potential for a high level of identity with the building, with residents knowing each other and thus creating strong social bonds. In complex buildings, towers are often used to form the corners of perimeter blocks and/or may serve as formal accents within the larger complex.

Slab

The most common form of housing massing is the slab, generally a rectangular form of building created by placing units side by side along a street. A slab can be made up of a block of rowhouses, or for a taller building, units stacked along a corridor, either double loaded (with units on both sides), single loaded (with units along on one side creating an access gallery, typically outdoors), or in a skip-stop formation (with two-story units along a single- or double-loaded corridor, thus avoiding the necessity of having a corridor on every level). In complex housing a housing slab may be used as a street edge, may serve as a landmark in the landscape, and may also be combined with other blocks or towers to create an implied courtyard. Sometimes, slabs are suspended from adjacent buildings. A perimeter block is made up of several slab masses that create the new form.

Perimeter Block

A common pattern for housing in European cities is an urban block comprising a courtyard at the center created by housing slabs around all or part of the exterior. In the Netherlands, these traditional courtyards, called *hofs*, or *hofjes*, are often accessible to the public during the day. Those created in the Middle Ages originated as agricultural plots where residents could grow food. Today the courtyard provides light, air, a view, sometimes shared social space, and/or access to gardens. Many of the complex housing projects are organized around courtyards. The courtyards vary in that some have exclusively public space, others are semi-private areas shared by residents in surrounding units, and still others include or are devoted exclusively to space associated with a living unit. But in most of the complex housing discussed here, the courtyards are secure and open only to residents, or function simply to provide light and ventilation.

Courtyards

The massing elements or perimeter blocks can enclose a courtyard, and fingers can generate courtyards with one open side. Courtyards have many different uses, including serving as a rain garden for water collection and recycling.

Access

When units are accessed from a courtyard, the courtyard may be primarily circulation space, or may combine circulation with adjacent units' exterior space.

Community

These courtyards are typically accessed by all residents or by a group of residents and are designed to encourage socialization. They may contain furniture and be directly entered from non-housing functions that relate to social purposes.

Garden

Often courtyards serve as gardens for dwellings at the base of the building. In some cases, only those dwellings have physical access to the space, although other units may enjoy visual access to these areas. In other situations, communal gardens exist beside those of individual units.

Play

Children's recreational needs are served by play courtyards.

Visual

Visual courtyards are intended to be used by residents exclusively for a pleasant view.

Lightwell

Sometimes courtyards serve solely for access to light and ventilation.

Parking

In urban settings, courtyards may be utilized for parking. When the open courts are very wide, they may include gardens adjacent to the parking. If they are narrower, decks or balconies may be located above the parking.

Commerce

Commercial courtyards, typically at ground level, provide access to businesses and therefore are open to the public.

Articulation of Form

The way that building elements and details are arranged can make a building either appear smaller or to appear interesting to the eye, thus to affect the perceived density. Complex housing projects employ a variety of articulation techniques and elements to fit into their context without overpowering the adjacent context or the person approaching the building. For instance, cantilevers may be used to create additional interior space without adding volume or barriers to the ground plane. Articulation of shapes and materials can be used to create a building that reads as a unified whole or as a building that reads as a set of parts, or that has elements of both readings.

Bridge, Gate, and Window

Openings in a slab may be employed in a variety of ways such as to allow light into a courtyard, to allow access into a space, or to generate a sense of connection between sides of a structure. In some cases these openings are in the middle of a wall, and in other cases they form an entry or as a passage through a building. Such reduction of the overall mass of a slab makes it appear more transparent and inviting.

Setbacks

Many designs employ setbacks at the top of the building and/or at the bottom. When used at the top, the setback makes the upper level or levels less visible from the street and thus reduces the apparent height. At the bottom of the building, setting back the lowest levels or two reduces the apparent connection to the ground, making the appearance less massive or even creating a sense of the building floating. Within the building plane, small setbacks such as loggias (see Outdoor Space) can create variety in the building plane, reducing monotony.

Cantilevers

Whether in the form of a slab cantilever or a balcony (see Outdoor Space), overhangs can create surprising and dramatic relationships or detailed visual play against a building wall. The shadows generated by overhangs tend to emphasize the three-dimensional aspect of what might otherwise seem a two-dimensional façade. Balconies are often used to create visual interest in a façade.

A building can communicate a more conservative attitude with a smooth building surface, a moderately conservative attitude with a repetitive and possibly symmetrical application of overhangs, or a dramatic effect with the use of asymmetry and a variety of cantilever shape and size.

Fenestration

Windows and doors may be grouped or used individually to create different effects. As just mentioned, symmetrical and repetitious application tends to make a more conservative appearance. The fenestration may be used to communicate different building functions. For example, wide and tall windows at the lowest level of a building may indicate commercial activity, while having many long ribbon windows projects an office activity. In housing, we can often read the location of a living room by a long window or glass door and balcony, a bedroom or bathroom by a shorter window. We may be able to identify a loggia because typically they have a several floor-to-ceiling windows area with a protective rail behind. A unit's windows may be grouped in a panel to express the identity of each dwelling. A building with a uniform or repetitive fenestration pattern tends to read as a whole, whereas a building with a variety of window patterns, depending upon how they are placed, may read as a having different parts.

Material and Color

A building with a consistent material and color tends to read as unified. Building façades may reflect or hide what is behind them. Materials and colors are often used to differentiate sections of a building or to reduce the apparent scale of a building by visually breaking it up into smaller sections that have a different texture or hue.

Typological Analysis—Housing Type and Unit Type

In this analysis, we address both housing types and unit types. Although the great emphasis of this discussion is on housing types, it is important to

Types of housing and unit

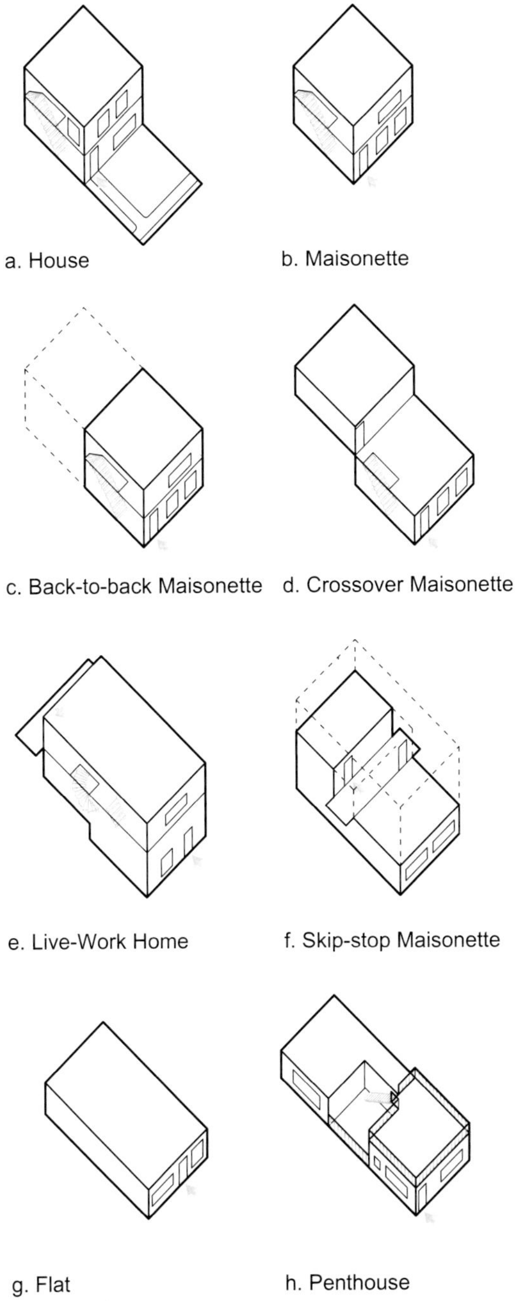

acknowledge that within a given housing type, there is an almost infinite variety of unit types that can be designed to respond to orientation, to size, to access to outdoor space, to room arrangement, to fenestration, and so forth. The standard approach to dense housing or *simple dense housing* is layered up to the required number of levels, and usually includes one housing type with as few as one or two unit types. In the Dutch design process, typically the urban designer (hired by the municipality or the developer) determines the set of housing types within a given block, and the architects design the unit types to fit. Given an identical set of housing types, each architect will likely create different unit types.

Here, housing type is defined both by form and by purpose. In terms of form, housing types are categorized by their arrangement (whether one story, flat, or multiple stories: rowhouse or maisonette), and by their type of access (direct access from the ground level, corridor access, vestibule or cluster access, or gallery access). In terms of purpose, they are defined by their use or by the type of resident they serve (penthouse, lightwell housing, live/work and atelier units, group home, assisted living, and guest flat). To analyze the form of the housing type, we first examine all of the units in terms of the number of levels and their access. To understand the diversity of purpose in terms of housing type, we add the special uses.

Formal Housing Type

Rowhouse and Maisonette

Rowhouse and maisonette units have more than one level with a connecting interior stair. They are differentiated by their mode of access, with the rowhouse being entered from the street and often having a garden at ground level, and the maisonette being entered from a corridor and typically encompassing a balcony or loggia. Because both types have more than one level, rowhouses and maisonettes provide a significant amount of spatial separation between living areas and sleeping areas, and in the case of live-work units, working spaces. Such separation makes these ideal units for families.

These dwellings resemble the traditional European single-family city house familiar to Dutch inhabitants: in shape, the rowhouse and maisonette are generally tall, narrow and deep, with access to light and air on front and back, and for end units, on the side as well. Their positioning provides orientation and cross-ventilation to at least two directions. Occasionally rowhouses and maisonettes are designed to have access only on one side. This limits orientation of windows to one direction, reducing the quality of light and ventilation. Stairs, in fact quite steep stairs by American standards, are very acceptable in the Dutch context, which is one reason maisonettes are quite common. Because they comprise at least two levels, architecturally these dwellings offer the designer a way to vary a façade so that every level is not the same.

Flat

For many residents, including the elderly and the physically handicapped, a dwelling with all the rooms on one level is necessary or highly desirable. Like maisonettes, when flats are arranged on a single-loaded corridor or gallery (gallery flat) they have access to light and ventilation from two directions permitting cross-ventilation and access to light across the day. However, flats on a double-loaded corridor, unless they are corner units, are limited to one orientation. Flats at the top of a building may have an outdoor space open to above (atrium flat). Among these projects, flats that enter from an elevator vestibule (vestibule or cluster flat) are typically corner units that have two orientations. The term flat is chosen rather than apartment because apartments are not always on one level, and flat is unambiguous.

Functional Housing Types

Live-Work Unit

Live-work units combine a dwelling with a workplace. Normally they are rowhouses or maisonettes with direct access to the street with the workspace at the ground plane and dwelling above. A special variety of live-work space is the atelier, which includes a workspace with extra ceiling height. Sometimes live-work units are not on the ground floor, and will not have direct access from the street.

Assisted Living

Assisted living dwellings provide independent living units for inhabitants, typically seniors or people with disabilities, in combination with social services. Such dwellings are usually clustered together within a project to achieve efficiency.

Group Home

Group homes in the Netherlands may be located within multi-family dwellings. These units are designed to provide individual bedrooms for several residents who share the living, kitchen, dining, toilet, and bathing spaces and are supervised by staff. Group homes may take the form of flats, rowhouses, or maisonettes. In some cases an adjacent dwelling for staff is provided as well. Since the construction of these projects, a trend has emerged to replace group homes with individual, dispersed, supervised dwelling units in such complexes.

Penthouse

Penthouse is selected as a unit type as it (1) has a special location—usually at the top of a building; (2) can have one or multiple levels; (3) is entered from a gallery, corridor, or vestibule; (4) incorporates a special type of outdoor space— the roof garden, terrace, or deck; and (5) usually serves upper-income residents. Since it is typically at the top of a building, and most often a luxury unit, the penthouse usually provides a generous amount of space, lots of fenestration, and often includes skylights and excellent views.

Unit Type

A unit type is a formal variation on a type of housing. There are many ways that dwellings may be configured, and here we try to encapsulate some of the organizational differences that are found in complex housing. Most of these unit types could be understood to form a subtype for several of the housing types.

Back-to-Back

Flats, maisonettes, and rowhouses can be positioned not only with shared sidewalls, but also back-to-back, with shared rear walls. The resulting units are tall, wide and shallow in depth, with access to light and ventilation only on one side. The disadvantage of such arrangements is that it does not permit cross-ventilation, and the spaces are all oriented in one direction, which may not be the best one. An unusual arrangement of maisonettes and rowhouses that can correct for this is a "cross-over" design (see "Cross-over"), which transforms units that would have been back-to-back to units that have access to light and ventilation from two orientations instead of one. Typically back-to-back units are on the ground level and thus rowhouses, because if used for maisonettes or flats, access would be difficult to arrange.

Cross-over

Cross-over units can take what might have begun as a back-to-back scheme and transform it. Instead of having the second floor of a unit on top of the first, the second floor of the unit is on top of the first floor of the unit behind it, and conversely, the adjacent unit has its second floor over the first floor of the initial unit. Two dwellings that would have been back-to-back, locate their second levels over the first level of the other, so that the bottom level faces one side of the building and the top level faces the other. A stair connects the two floors. This permits orientation in two directions and cross-ventilation with only slightly less efficiency than the back-to-back scheme. Units can also cross-over with the adjacent unit, but this doesn't correct the problem of a single orientation.

Skip-Stop

In the skip-stop arrangement, originally developed by Le Corbusier for his Unité d'Habitation housing scheme in Marseilles, a double-loaded corridor provides access to paired two-story units. One of these includes an internal stair to the full level above the corridor and the other unit has a stair to the full level below, thus reducing the number of corridors and elevator stops by more than 50%, while increasing the number of units on each corridor. In this arrangement, unit entrances are usually on opposite sides of the double-loaded hallway. Both units typically have some living areas at the corridor level, with the remainder on the floor through living space above or below the entrance level. Thus, both units have orientation to both sides and have cross-ventilation. Alternate arrangements may create split-level units or gallery access. In any case, the structure resulting from the arrangement is a block of side-by-side dwellings that forms a wall or a bridge several stories tall.

The disadvantage of the typical split-stop structure is the double-loaded corridor that is hard to supervise and lacks natural light and ventilation. Sometimes the split-level pattern is used to generate extra tall corridors to create a greater sense of openness, although the split-level approach is not used in any of the Dutch examples studied.

Lightwell and Atrium

Lightwell and atrium units use roof openings to allow light and ventilation into the center of a unit in addition to having openings through the horizontal walls. The lightwell enables the creation of a deep unit since interior spaces can be lit and ventilated through the well, although a small well may compromise the ventilation and light quality. When the lightwell is large enough and designed to be used as a habitable outdoor space, it is called an atrium.

Access

Here we discuss the various modes of access from the ground to the dwelling, both vertical and horizontal.

Elevator and Stair

Vertical access organizes the housing block. The access elevators, access stairs, and exit stairs, usually designed as a single element, are strategically located for fire exiting, to maximize the number of units each serves, and to support the various types of horizontal access to the units provided in the project. The placement is particularly challenging when there are various types of horizontal access that need to be coordinated with each other.

Ground

Direct access to one's own unit from a street or courtyard is often considered an ideal arrangement, although entering off a busy street can sometimes be a problem for security or sound transmission. These problems are often resolved by providing a buffer space, such as a level change or small garden area. For families, it can be a decided advantage when a ground level unit has its own garden behind the dwelling, although this is not always the case. Sometimes a rowhouse has parking on the ground level behind the

Types of access

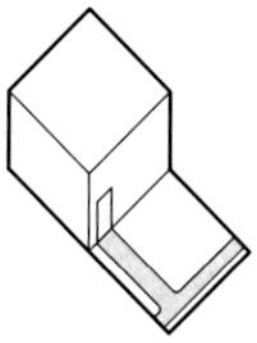

a. Ground Level

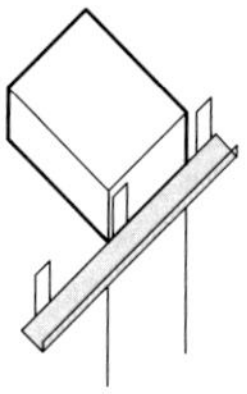

b. Gallery / Single-loaded Corridor

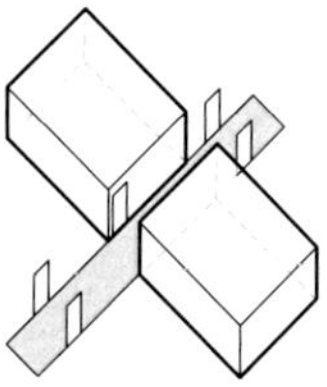

c. Double-loaded Corridor

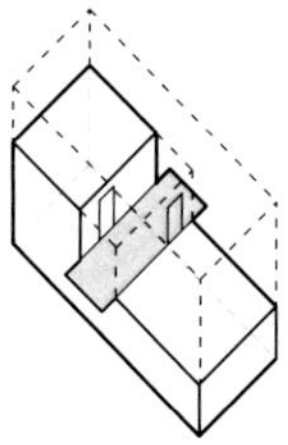

d. Skip-stop Double-loaded Corridor

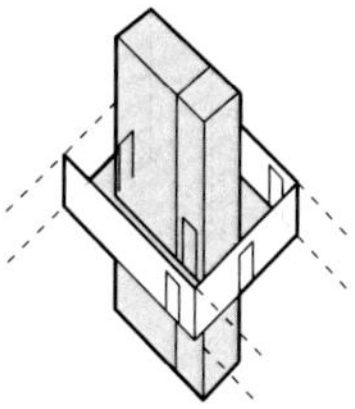

e. Vestibule / Cluster

unit with a deck above for outdoor space. For live-work units, street access provides business opportunities.

In many Dutch cities, there is a tradition of doors to many units from the ground level. Within a single building, there may be many walk-up apartments each with its front door on the street and its own staircase that may go up two, three, or even four levels to the unit. But today when elevators are more available, although people will walk up one or two levels, access stairs that go up more than two levels are not considered desirable.

As noted elsewhere, another important consideration in relation to ground access is the effect it has on the neighborhood. When there are many entrances directly on the street, it makes the street a livelier place than when there is a single entrance used by everyone in a housing block.

Gallery and Deck

In the Netherlands, the single-loaded gallery open to the air is often used as a solution for circulation in large buildings, in part because it provides just the separation between front door and fire stair required by fire code. At the same time, it provides a sense of being a public street, since the gallery is open to the air and to the entries of the adjacent dwellings. Additionally, the units on the open corridor or deck (surfaced in wood) have access to light and air on both the circulation and street sides, thus meeting the Dutch requirements for high levels of light and ventilation. In courtyard buildings, open corridors provide visibility of access routes across the open space, useful for security and socialization both from windows of units along the corridor and from those on the opposite side of the courtyard.

A disadvantage exists when there are so many units along very long galleries that it impedes the recognition of neighbors, so necessary to supporting an informal social network. And of course, access to the outdoors has both benefits in beautiful weather and liabilities in inclement weather, especially when there is snow and ice.

Another disadvantage of the single-loaded corridor, gallery, or deck is that several layers of galleries can make a monotonous façade on the access side, although that façade is typically invisible from the street. Combining flats and maisonettes, and limiting the length of the gallery, can create a façade with more variety.

Single-loaded galleries, or even the rare double-loaded outdoor galleries may be widened to create more shared circulation space for socialization in the form of access decks. In cold climates such as the Netherlands, these are especially effective when oriented to the sun and can double as outdoor space for the adjacent units.

Corridor or Double-Loaded Hallway

Interior halls with dwelling units on both sides are an efficient way to share elevators and other resources among a large number of people. Furthermore, they provide protection from the weather, but often without natural light.

In the 1950s, when a tall building was desired but the cost of an elevator was very high, in the Netherlands and all over the world, buildings with many dwelling units were often designed with many units on a corridor, all connected to an elevator core. These long circulation spaces, especially when double-loaded, were difficult to supervise visually and often became insecure and unsafe. Problems of security can be reduced by limiting the number of units on a corridor, by having openings into the corridor to provide visual access from inside the units, by having sufficient artificial light, and/or by providing natural light and ventilation.

In addition, when the units along the double-loaded corridor are flats, even when small in number, unless they are on a corner or have a very special design, they will have access to just one orientation, resulting in no cross-ventilation and limited access to sunlight, especially if on the north side without access to sun. In a climate with hot summers and cold winters, this can be a problem.

As mentioned earlier, in the Netherlands fire codes address smoke inhalation, so that when double-loaded corridors are used, a balcony or vestibule open to the outdoors is required between the corridor and the fire stair.

Vestibule or Cluster Hallway

The vestibule access involves a cluster of units with doors so close to the elevator vestibule and stair that there is no need for a corridor. In the Netherlands, this arrangement includes an outdoor space between the vestibule and the exit stair. Vestibule access is occasionally used for a skip-stop arrangement.

Having a limited number of units on each level with all unit doors visible to each other and visible from the elevator vestibule is advantageous, creating a high likelihood of residents being familiar with each other, thereby promoting community and security.

Outdoor Space

Garden and Rooftop Garden

One does not usually associate multi-family housing with having a front door or a garden, but several of the complex housing projects have rowhouse units at the base of the building that open onto gardens or courtyards, and penthouse units with roof gardens. Although similar to a garden in a private house, in complex housing, because the rowhouse gardens and some penthouse gardens are visible from above or from the side, privacy is reduced. Nevertheless, having access

Types of outdoor space

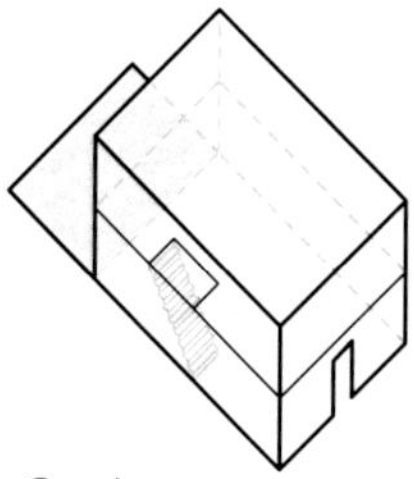

a. Garden

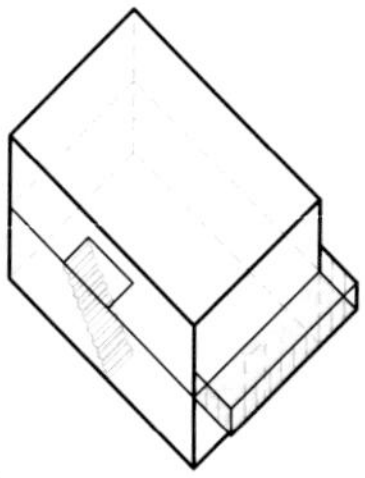

b. Balcony

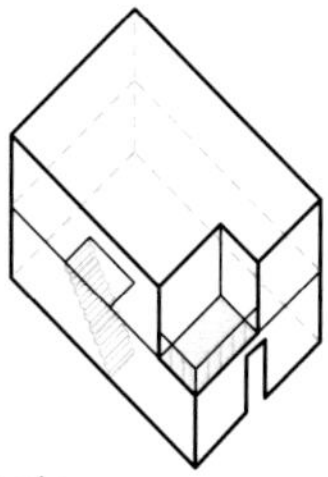

c. Loggia

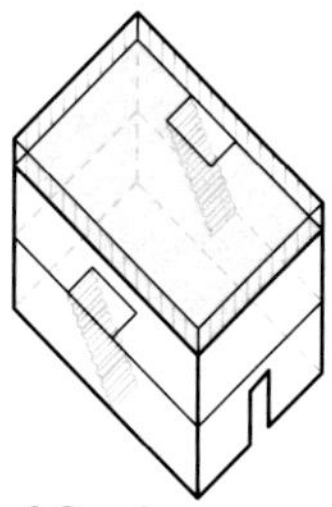

d. Roof Garden

to a garden can be an important asset to a dwelling unit, especially in the city.

Increasingly, roofs are used as outdoor living spaces. This is simplest when the roof is adjacent to the living area, but several dwellings in this study provide garden space on top of the dwelling accessible from a set of stairs. Being at a high location, such spaces frequently have excellent views of the surroundings while exposed to sun and wind.

Balcony

Although the situations may have changed since then, during most of the period of the VINEX construction almost every unit was provided with some outdoor space. Balconies are a common solution, with many sizes, shapes, and materials. Sometimes open to the air, enclosed simply by a metal or wood grill, sometimes enclosed by glass that can protect from the wind, and other times enclosed with a solid wall for privacy, balconies are frequently used by architects to enliven the façade. Of course good solar orientation is essential to a balcony that works, so balconies are most often found on the east, south or west façades.

Loggia

In contrast to balconies, which are generally cantilevered from the building wall, loggias are areas open to the out-of-doors, usually covered and cut out of the building wall. Orientation to the sun is also highly desirable in such spaces. Within the building skin, these spaces can be very protected and private, or when made of glass, are like conservatories or indoor-outdoor rooms. Often the exterior wall of a glass loggia will have doors that open up like an accordion so that the entire area is out-of-doors. The wall between the loggia and the dwelling is typically of glass as well. In the Netherlands, loggias with soundproofed glass are commonly used to provide outdoor space when buildings are located near railroad lines or busy streets. They are also used to create a protected outside territory in windy areas. When oriented to the sun, loggias are usable on a sunny day in every season of the year, sometimes with the doors open, at others with the doors closed.

Organization, Syntactical Structure, and Social Principles

The differences and similarities in organizational structure are very important to understanding the principles used to design complex Dutch housing. Distinct approaches to density, scale and layout affect the perceived character and perceived density of the project. As mentioned earlier, Jane Jacobs suggests that when densities become too high, the result is standardization.[38] Because this research involved no systematic observation of behavior and only minimal feedback from residents of the different projects, the conclusions related to social principles are primarily based on interviews of designers and developers and from examination of the architecture. Embedded in the designs themselves are implicit ideas about the numbers of people that can share a hall and can use an elevator, and about such things as where visibility of other residents is important and where it is not as important. The potential for social interaction is inferred from the individual design decisions and the organization and structure.

Number of Dwellings

Size can be designated by floor area, by the number of units, and by the number of circulation spaces in the syntax diagram.

Subdivisions within the Projects

By looking at the floor plans and the gamma analysis diagrams we can identify how the plan is broken down in terms of access points, corridors, and vertical circulation. We can differentiate between spaces that are entered directly from the street, and those that require entering the building complex. The gamma diagram threads show how the internal spaces are structured relative to the visitor.

A key attribute of the structure largely visible in the gamma diagrams is the circulation threads or circulation systems. For this project there are three types of circulation systems: (1) directly from a street (with each street defined as a separate, invisible thread), (2) through a courtyard (each courtyard a separate thread), and (3) from an elevator system. Threads with clear, fan-shaped arrangements are called *collected*. If the threads have few spaces along them, the spaces are *dispersed*. The threads can be similar to each other with repetitive patterns or may contrast. The threads can be independent of each other or highly connected. The threads can be *clearly organized* or more *loosely structured*. The threads may be deeper or shallower. There may be many threads or just a few, compared to the number of units overall. These differences suggest degrees of control and access as well as the potential for social cohesion.

Mix

The mix of functions is evident from the floor plans, due to the difference in configuration of dwellings and other purposes. We can understand the mix of unit types by examining the plans, but the mix of income levels and units for sale or purchase is not evident from the plans or diagrams, and important to consider in the overall structure of the space.

Application

In the next eight chapters the examples of Dutch complex housing will be studied using these typological categories. As we examine them, lessons about their design will become evident and contribute to the conclusions about design principles and replicability.

Notes

1. Robert K. Yin, *Case study research: Design and methods* (Thousand Oaks, CA: Sage, 1994), 13.
2. Vitruvius Pollio, *Vitruvius: The ten books on architecture*, trans. Morris Hicky Morgan (New York: Dover, 1960).
3. Leon B. Alberti, *The architecture of Leon Batista Alberti in ten books* (London: Edward Owen, 1755).
4. Andrea Palladio, *The four books of architecture*, ed. Adolf K. Placzek (New York: Dover, 1965).
5. Robert Cervero, and Erick Guerra, "Urban densities and transit: A multi-dimensional perspective," Working Paper UCB-ITS-VWP-2011–6, Berkeley, CA: Institute of Transportation Studies, 2011, 15.
6. Meta Berghauser-Pont and Per Haupt, *Spacematrix: Space, density and urban form* (Rotterdam: NAi, 2010), 79–204.
7. Ibid., 81.
8. The decision to use the housing footprint rather than the site dimensions allows for a clearer comparison because the width of a street and the landscaped area around a building are not uniform across projects. Furthermore it is difficult to get reliable site information. Therefore floor plan data is more reliable and accurate. However, this calculation method creates higher densities than using the site parameters, because the building footprint is generally smaller than the site data. Because of the unreliability of our site data we were not able to accurately calculate the difference, but it seemed to range from 15% to over 50%. When comparing these data to other density data, this difference needs to be taken into account.
9. Robert Edward Mitchell, "Some social implications of high density housing," *American Sociological Review* 36 (February, 1971), 18.

10. Daniel Stokols, "On the distinction between density and crowding: Some implications for future research," *Psychological Review* 79, no. 3 (1972), 275–277.
11. Amos Rapoport, "Toward a redefinition of density," *Environment and Behavior* 7, no. 2 (1975), 133–158.
12. Berghauser-Pont and Haupt, *Spacematrix*, 229.
13. Jane Jacobs, *The death and life of great American cities* (New York: Random House, 1961), 209.
14. Ibid., 209.
15. Ibid., 211–217.
16. Javier Mozas and Aurora Fernández Per, *Densidad: nueva vivienda colectiva* [Density: New collective housing] (Vitoria-Gasteiz: a+t Ediciones, 2006), 15.
17. Ibid., 15.
18. Bill Hillier and Julienne Hanson, *The social logic of space* (Cambridge: Cambridge University Press, 1984), 143–175.
19. Bill Hillier, Adrian Leaman, Paul Stansall, and Michael Bedford, "Space syntax," *Environment and Planning B: Planning and Design* 3, no. 2 (1976): 147–185.
20. Martin Krampen, *Meaning in the urban environment* (London: Pion, 1979).
21. Donald Preziosi, *The semiotics of the built environment: An introduction to architectonic analysis* (Bloomington: Indiana University Press, 1980).
22. For example, see: Hillier and Hanson, *The social logic of space*; John Peponis, Catherine Ross, and Mahbub Rashid, "The structure of urban space, movement and co-presence: The case of Atlanta," *Geoforum* 28, no. 3 (1997): 341–358; and Alan Penn and Alasdair Turner, "Space syntax based agent simulation," in *Pedestrian and evacuation dynamics*, edited by Michael Schreckenberg and Som Deo Sharma (Berlin: Springer-Verlag, 2002), 99–114.
23. Thomas A. Markus, *Buildings & power: Freedom and control in the origin of modern building types* (London: Routledge, 1993).
24. Julienne Hanson, *Decoding homes and houses* (Cambridge: Cambridge University Press, 2003).
25. Jean-Nicolas-Louis Durand, *Recueil et parallèle des édifices de tout genre, anciens et modernes, remarquables par leur beauté, par leur grandeur ou par leur singularité, et dessinés sur une même echelle* (Paris: Chez la auteur, 1801).
26. Roger H. Clark and Michael Pause, *Analysis of precedent: An investigation of elements, relationships and ordering ideas, in the work of eight architects* (Raleigh: Student Publication of the School of Design, North Carolina State University, 1979).
27. Anne Vernez Moudon, *Built for change: Neighborhood architecture in San Francisco* (Cambridge, MA: MIT Press, 1986).
28. Susanne Komossa, Han Meyer, Max Risselada, Sabien Thomaes, and Nynke Jutten, *Atlas of the Dutch urban block* (Bussum: Thoth, 2005).
29. Mozas and Per, *Densidad*.
30. Albert Ferré, Tihamér H. Salji, and Actar, *Total housing: Alternatives to urban sprawl* (Barcelona: Actar, 2010).
31. Peter Ebner, *Typology+: Innovative residential architecture* (Basel: Birkhäuser, 2010).
32. Hillier and Hanson, *The social logic of space*.
33. Markus, *Buildings & power*.
34. Several of the architects interviewed mentioned how they employed typology in their designs, including Roelf Steenhuis (Architect, Steenhuis-Bukman), Interview, Delft, March 8, 2010; and Frits Van Dongen (Architect, van Dongen-Kususch), Interview, Amsterdam, March 2, 2015.
35. Yin, *Case study research*, 34.
36. Julia W. Robinson, "The question of type," in *Ordering space: Types in architecture and design*, edited by Karen A. Franck and Lynda H. Schneekloth (New York: Van Nostrand Reinhold, 1994), 179–192.
37. For example, Roger Sherwood, "HousingPrototypes.org: Multi-Family Housing," www.housingprototypes.org/, 2002–2016. Mozas and Per, *Densidad*, 15–35; and Ebner, *Typology+*, 20–29, 111–121, 244–253, 322–327.
38. Jacobs, *The death and life*, 209.

2 De Muzen

De Muzen south façade

Location of De Muzen in the Netherlands

Location of De Muzen in Almere

Location of De Muzen within its neighborhood

De Muzen is located in Almere, a polder city within a half hour train ride east of Amsterdam. The land that forms Almere was drained as a polder from the Zuiderzee (South Sea) in the late 1950s and early 1960s. The first housing projects were built there in 1978, and today, about 40 years later, it has a population of almost 200,000. Assuming households of 2.5 people, this means that approximately 2,000 housing units have been built each year. Because the city continues to grow (and anticipates a population of 300,000 by 2050), this is an ongoing feat of extraordinary planning. The city has developed in several large sectors, one of which is Almere Stad West, which includes the Muziekwijk neighborhood where De Muzen is located.[1]

De Muzen is chosen as an early example of complex housing, as well as for its status as a landmark in a suburban area, even though it lacks the diversity of residents found in later examples. The project was designed and built in 1997 as housing for elderly for rent and purchase by the architecture firm Atelier Pro Architekten (Leon Thier and Noud te Riele with Arjan Oosterhof) and the landscape architecture firm Copijn (Jorn Copijn and Lia Copijn-Schukking). De Muzen's construction was the finishing touch on the Muziekwijk Noord neighborhood in Almere. All the surrounding housing had been built, and this last project was to be a long, semi-circular building that framed the end of the long and thin Muziekpark at the center of the Muziekwijk neighborhood (see photo page 50).[2] Its landmark visibility gave it a high priority for the city and developer, Groene Stad Almere (now De Alliantie Flevoland, Ymere).

The density of De Muzen, when calculated using the entire site area, reflects the suburban location at 80 units per hectare (30 units per acre), the lowest of the group. However, when calculated using the footprint alone, as if it were built instead on a tight city site, the density

Aerial view of De Muzen showing its prominent location at the north end of Muziekpark, the parking area to the north of the complex, and the low-rise housing surrounding it © Hans Brons RoVorm

Competition sketch of De Muzen by Noud te Riele © Noud te Riele

Model of the city proposal for the De Muzen site (model-maker Frank Heerema) © Peter de Ruig

below and page 51).[3] According to architect Noud te Riele,

> We needed a lot of time to convince [the municipality] of the project. But the quality of the project was obvious. . . .
> . . . If you are the builder, it is quite a statement to make such a social housing project. And if you win a competition, the municipality can say what it wants, but you are in a very strong position. That is a good way to start a project.[4]

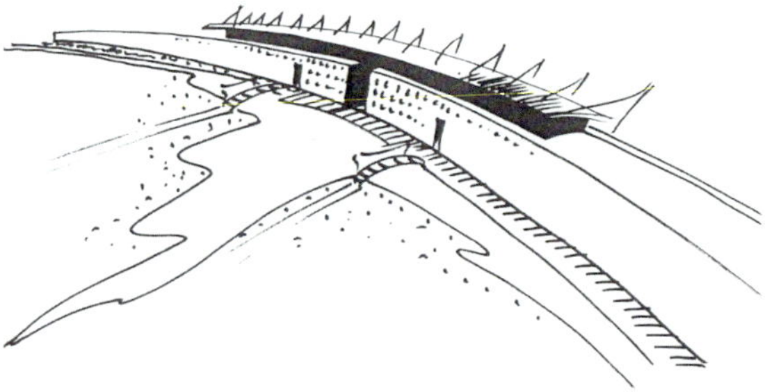

is much higher at 340 units per hectare (140 units per acre). When compared to the densities of the other projects calculated in a similar manner, De Muzen turns out to be one of the most dense projects, second only to Silodam. In other words, if a project similar in design were to be built on an urban site where parking is under the building rather than on the adjacent site, the net density per hectare might well be four times the net density of this project on its present site.

Site, Context, and Building Functions

Atelier Pro won a competition for the De Muzen project and proceeded with the design in cooperation with developer Groene Stad Almere. The winning proposal did not match the competition requirements. Almere's urban design scheme was semicircular (see photo to the right), had separate blocks for housing at each end, and was white in color like the nearby housing, while the Atelier Pro proposal had only a slight curve, was one long connected block, had an interior atrium, was located on the park site, and was colored red. Finally the proposed building was a complex design (see photos

Model of the Atelier Pro competition proposal for the De Muzen site (model-maker Frank Heerema) © Raoul Suemondt

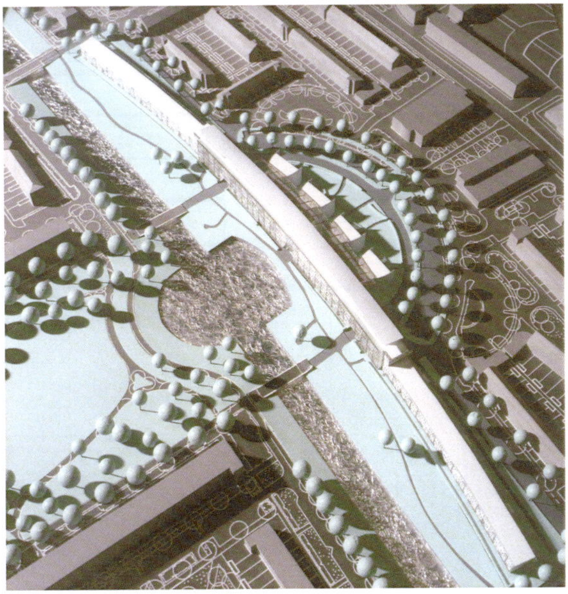

After initial discussions with the neighborhood, te Riele was also concerned about the impact of having a necessarily tall building on the site that had previously been open land. Furthermore, the cost turned out to be too high to move forward exactly as planned. He proposed (1) enlarging the canal to create a body of water in front of the project, (2) creating one long connected structure instead of having two separate groups of low buildings at the end of the main block, (3) keeping the atrium, but reducing its size, and (4) adding the tulip structures on the west side to create more units.[5] But the city and the neighbors had to be convinced. The developer David Wolterbeek describes one meeting with the neighbors:

> The city had made a deep curve. The architect suggested a less strong curve. In fact it was a brilliant idea of Noud. He made a presentation with a wooden stick, and four lamps he called tulips. Everybody thought a good idea to have two fronts. The curve alleviated the sense of a long gallery. And the less deep curve made it possible to have more space in front of the houses. . . . For Noud, it was not a bad idea to get the neighbors on his side.[6]

The project was designed as housing for the elderly with 80 social housing units, 75 moderate-income units for purchase, and 10 penthouse units for the upper-income market.[7] Additionally it included two guest apartments, a communal space, and the atrium.[8] The units associated with the interior courtyard included those for assisted living.[9] But as the project evolved, the project budget and interests of the city drove the rental social housing to become rental for middle-income residents, while the housing for purchase remained aimed at the original economic group. Furthermore, due to market forces at the time of sale, a number of dwellings were sold on the open market to younger people.[10]

> We wanted to make an atrium the elderly people could use year round. And we really wanted to make a palace for elderly people where you can grow older in a nice way. It's easy to attract health care. The building is very close to a health center and a shopping area. Also public transport is very close and well organized.[11]

The parking is entirely outdoors, placed in a well-landscaped area between the project and the nearby housing. Graded lower than the housing to reduce its visual impact, the parking lot is arranged in a semi-circle that follows the footprint of the original city proposal. Across the street from De Muzen beyond the parking lot are conveniently located a health clinic, a supermarket, and some shops. The rail station and shopping mall are both within walking distance.[12]

Organization, Appearance, Form, and Massing

At 360 meters or almost a quarter mile, De Muzen is an unusually long, curved salmon-colored housing block, its powerful presence somewhat reminiscent of the housing

Diagram of the elements of De Muzen: rowhouses, park-facing flats, penthouses, atrium, and tulip clusters

The main block of De Muzen in 1997 © Ger van der Vlugt

De Muzen section

The garden atrium with rowhouse entrances

at Bath, England (more obviously in Almere's original urban design). The complex has two sides, the long one facing the park, comprising two stories of rowhouses with gallery flats above, and the shorter one facing the housing block consisting of the four tulip-shaped clusters of flats and maisonettes, between which is an interior garden courtyard overlooked by galleries. Inside, the garden atrium houses the vertical circulation and shared facilities including a two-story meeting and community space.[13]

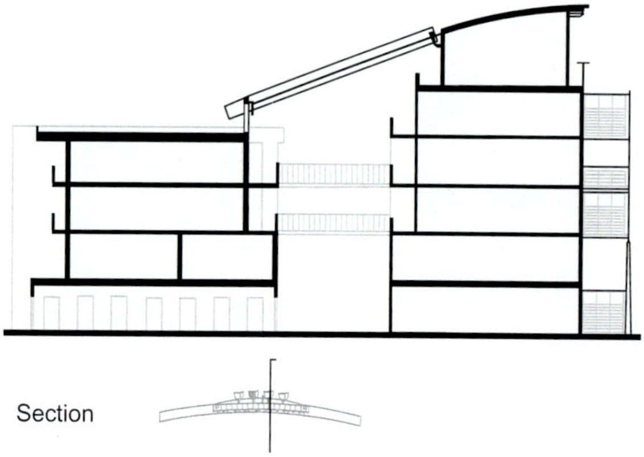

The architect noted "We tried hard to harmonize the buildings with the surroundings in terms of height."[14] De Muzen is surrounded by the low-rise, mostly single-family rowhousing typical of a Dutch suburban area like

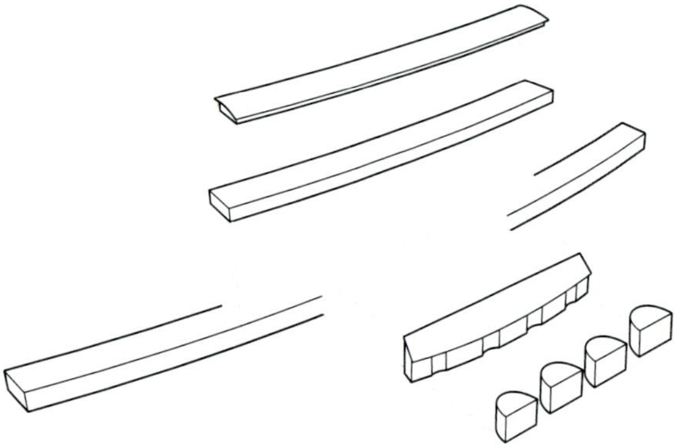

De Muzen floor plans

The building steps down near the bicycle and pedestrian gateway

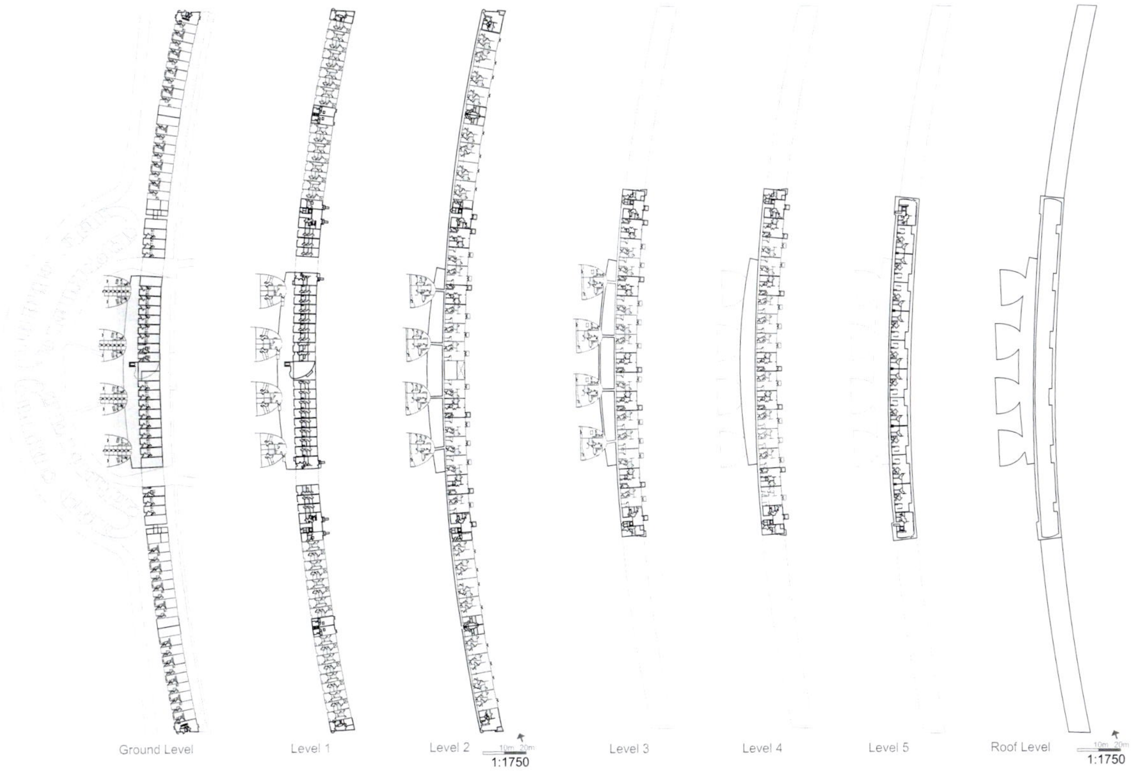

Almere. The design responds to the context by having the long block step down to three stories at the two wings, and by having the clusters four stories in height. Additionally, a link between the neighborhood and the park is created at the connection between the wings and the central structure where the building is pierced on both sides by a bicycle and pedestrian path.

One advantage of the long design is that almost all the dwellings in the project face the park. The dwellings in the tulip clusters are the only units that do not. In addition, employing rowhouses at the base of the structure accommodated neighbors' desires for no bedrooms on the ground floor. Every unit at De Muzen has an outdoor

The park façade block dips in the middle to create an extra level for the community space. Its pattern of balconies reflects the rhythm of the unit bays.

View into the community space shows the link to the park

space; the rowhouses have gardens or decks, and the flats and maisonettes have balconies. Storage areas large enough to include bicycles are dispersed at the base of the complex in about nine different places, so that all units are relatively close to a designated storage area. To create accessibility for the community space to both the interior courtyard and the park a floor below, the building dips down at the midpoint. Thus, the four rowhouses adjacent to the community space have decks rather than balconies.

Constructed as what the architect described as "a concrete tunnel structure," with structural bays creating an architectural grid, the building is finished with prefabricated insulated concrete and wood panels.[15] The strategic placement of three different unit widths and heights creates a façade with a syncopated rhythm, three against two against one. The line of rowhouses at the bottom are one bay wide and two units high, the flats on the next three floors are two bays wide and one level high, and the line of penthouses at the top are three bays in width and one level high.[16] The design incorporates a steady pattern of rowhouses lining the bottom and penthouses lining the top, and irregular placement of flats in between. Because the balconies for the flats are accessed uniformly from the living rooms, varying the placement of the living rooms in left or right bays permits the designer to create an asymmetric play of the white balconies that teases the eye against the red-pink concrete wall.

The color of the wood panels has changed over time, creating different effects. The panel color on the penthouse level has always been darker, further reinforcing the building's horizontality, although at some periods it has been much darker than at others. On the west side, the 2007 photograph of the darker colored clusters and main entrance seems overly somber when compared to the lighter color of the photograph taken in 2015 (see photos to the right).

The building is crowned by the penthouses at the top level. The distinctive curve of the roof reflects the arc of the building footprint and reinforces the building identity. At both sides of the penthouse level, the large setback that provides terraces for the end penthouse units reinforces the stepped effect of the lower wings. The shadow created from the roof overhang and the setback at the park side emphasizes the horizontal crown. Strengthening this effect are the brown wood panels, darker than the building. The recess along the park façade at the top level serves as an emergency fire exit for the penthouse units and varies in depth to create good-sized balconies.

This complexity of design was difficult to achieve. A small architect's fee did not allow for the development of the necessary architectural detailing. But the commitment of the city, developer, and architect to the project and the unusually strong cooperation between architect and developer allowed the project to be developed, as indicated in these comments by the architect.

> We worked together with the client for a long time on all the technical details for the plan. It was very hard to keep our ambitions high during the whole process . . .

North entrance in 2007 showing the darker wood panels

North entrance in 2015 showing lighter wood panels

> [The developer] just did a marvelous job. Of course the client was a very professional man. He [David Wolterbeek] changed all the details to make it work. We couldn't afford to do it. We had a low fee. At a certain moment everyone wanted to make this dream.[17]

This unusually strong relationship had to do with the leadership at Groene Stad Almere as described by David Wolterbeek.

> Jan de Vletter who was the director, was not a common type of director. He made a study of how to be a good client. The client

should know what he should ask and what he wants himself and how to combine these different interests. He wanted to be a good Opdrachtgever (commissioning company—the one who places the order) with a good spirit. When you are capable to define what you want and have a good vision for the community, all the things come together, you get a good product. . . . knowing that my boss would support making the right decisions. . . . This was a very special project.[18]

De Muzen unit plans

a. Row House

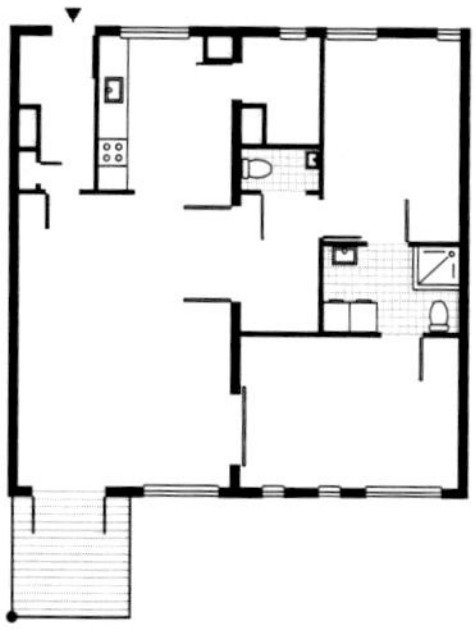

b. East-facing Flat

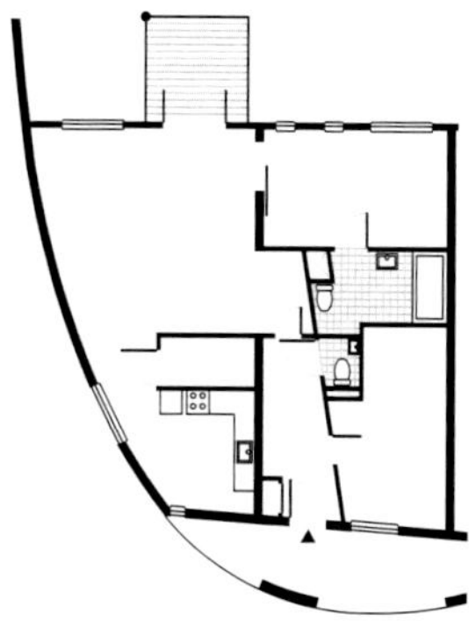

c. Tulip Flat

Housing and Unit Typology

While the original design was intended to include social housing, as the project developed and the circumstances of housing corporations changed, the social housing was eliminated in favor of middle-class rentals, although the housing for sale remained for middle- and upper-income households, primarily elderly. Accommodating the change required a slight increase in the housing bay from 4.80 to 5.10 meters to allow for well-designed staircases. The housing types are predominantly rowhouses and flats, with one maisonette and one guest unit in each of the four tulip clusters. According to the architect, there are fifteen different unit types, responding to different locations in the plan.[19] Here we discuss the predominant types: the rowhouse, the east-facing flat in the main block, the tulip flats, the tulip maisonettes and the penthouse.

Rowhouse

The rowhouses that form the base of the complex are one bay wide and designed with two or three bedrooms. The basic design is close to the traditional Dutch layout with a separate entry opening to the living room, staircase and WC, except that one reaches the kitchen via the living room instead of from the entry. Several of these dwellings have a two-story space in the living room, as drawn here. The staircase is either a straight single run across the unit or, where there is the open space, a single turning run along the kitchen wall (too steep for US

West wing: rowhouse entrances with gallery flats above

standards). Upstairs are the bath, two bedrooms, and either two storage areas or a storage area and opening to below. As mentioned earlier, the rowhouses uniformly have living areas on the ground floor with bedrooms above, so that no one in the complex sleeps on the ground level. The rowhouses in the wings have their own direct entry from the street, while those in the main block, both east-facing and tulip cluster, enter from the interior garden. All of the rowhouses have a garden or deck accessible directly from the living room.

Rowhouse living room: two-story space and garden beyond

East-Facing Flat

The east-facing gallery flats are not the traditional Dutch flat layout in that all the rooms are not accessed through the entry corridor (see plan b, page 56). The flats in the main block have living rooms and balconies that face east toward the park. These two-bay units have the entry and living area in one bay, and the bedrooms and bathrooms in the other. The master bedroom (facing the park) is reached directly from the living room, while the second bedroom (facing the gallery) is accessed via a secondary corridor space that also connects to the WC.

Atrium showing entrances to tulip flats on the top level gallery right, and to maisonettes from atrium staircases, below right

The bathroom has doors to both bedrooms. In this unit, the kitchen, having walls on three sides, is well defined and connected with the living space. These units have a square balcony accessible from the living area large enough for a table and chairs.

Tulip Flat

The tulip flats enter from the top level gallery that bridges the atrium or from stairs in the atrium (see plan c on page 56). The flats reflect the tulip shape, with curved kitchen and living room wall. The plan is nontraditional in that the entry corridor only controls access to the second bedroom, the WC, and the living room. To get to the kitchen and master bedroom, you need to pass through the living room. Again the bathroom is connected to both bedrooms. The west-facing living room opens to the large balcony, which has a neighborhood view.

Tulip Maisonette

The tulip maisonettes enter from the atrium or from the top level gallery. The narrow units have an entry that proceeds to a kitchen that opens onto the living room. Two bedrooms and one bathroom are up or down one level from the living area, reached from a stair located between the kitchen and living area. The west-facing living room opens to the balcony or garden.

Guest Unit

Noting that the apartments are small, the architect said, "We also made places for the children to visit."[20] These two one-room-plus-bath units are squeezed in between the two tulip maisonettes at the level above the atrium and reached by stairs. Their windows overlook the atrium.

Penthouse

Although the twelve 3-bay-wide penthouses are wider than the units below, they are less deep to accommodate the top-level setback and balcony. The extra width provides a very large living area. The plan has an untraditional layout for a Dutch dwelling. An entry hall with coat closet and WC leads to the living room area, and through one end of that space either to the dining area and kitchen, or to a small corridor area that links to the bedrooms and bathroom. Facing the park, the large master bedroom has glass doors onto the living room. The smaller bedroom window faces the access gallery. The outside spaces are reached from the dining room. Although the majority of penthouses have spacious balconies that face the park, the end units have an extra-large terrace on the side.

Penthouse dwellings: corner unit with side terrace, middle unit with park-facing Balcony

De Muzen space syntax gamma diagram

Syntactical Structure

As mentioned earlier, the rowhouses at the base of De Muzen enter either directly from the street or via the atrium. Most of the remaining units enter from galleries adjacent to the two main elevator cores, either exterior galleries as on the third level of the two wings, or galleries in the atrium courtyard. The exceptions are the two guest units and maisonettes and flats that enter from the atrium via stairs.

Despite having 167 units, which makes De Muzen the third largest of the projects studied, its syntactical structure shows only two threads related to its two elevator cores. The interconnected structure is similar to De Opgang, also primarily a gallery-accessed building, but which has half the number of units. The two fans at the base of the diagram represent the approximate one-third of De Muzen's units that are rowhouses entering directly from the street or atrium. The fans above are all units that share a gallery. The two fans on the exterior of the diagram represent the twelve units on each side of the atrium with an outdoor gallery. The remaining fans show the relationship of the dwellings accessible from the galleries around the atrium. The largest fan has twenty-five dwellings, which if divided by two makes no more than twelve units per elevator core on each level, a number that works well for social structure. All of the gallery dwellings are reached from the two elevator cores.

Because of the rowhouses' relation to the street and the three entrances to the atrium, the project is highly externally connected. It is also highly connected internally with the elevator cores linked to each other through the galleries. The threads show some complexity

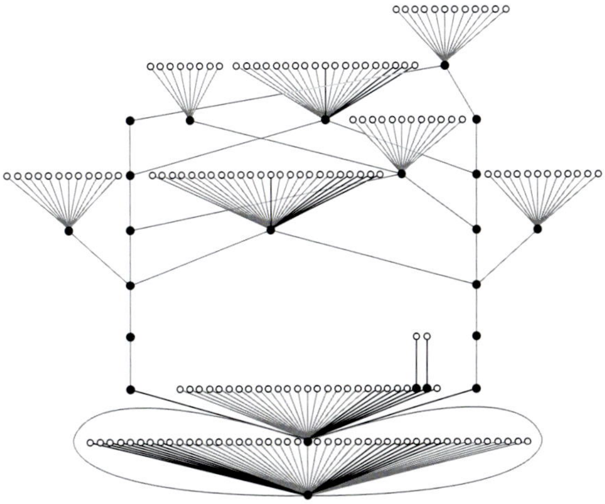

Atrium showing galleries and bridge from main gallery on the left to tulip flats on the right

and diversity of arrangement due to the unusual way that the tulip flats are accessed with bridges across the atrium.

Social Principles

Having an interior atrium in a project for the elderly is advantageous not only in providing a space with plants in the winter, but it creates a strong identity and social center for the project. This is reinforced by having entrances, galleries, and community spaces within it, where people can see each other on a regular basis. Also attention was paid to create pleasant circulation space in the atrium and galleries, including the use of wood on the walls and special lighting.[21]

By the time the project was built, the breadth of income levels of residents was limited due to the budget constraints and change in rules for housing corporations, but the age of residents was expanded to include more than just elderly. A real estate advertisement from Allantie indicates that now the population at De Muzen is limited to residents over age 35 without children.[22] But the provision of rowhouses, flats, and penthouses provides for a diversity of lifestyles, and the original design for the elderly effectively accommodates a wide range of ability. "We wanted the elderly people to live in the wings at first and then move to the center where there is more care."[23]

Lessons and Conclusions

De Muzen's powerful presence continues to exert its influence as an urban landmark. According to author Jaap van der Lans, the project plays an important role in the city.

> De Muzen . . . [due] to its distinctive character, its distinctive allure, gives the surrounding area an identity. People refer to it often to explain where they live. "Just behind De Muzen," for example.[24]

The project illustrates the impact of a project so well designed that it inspires the city, the developer, and the architect to build it in the best way possible. The importance of having the team work together cannot be overestimated. This was not an ordinary project. The care and diligence of the developer and city to support the proper construction assured the practical success of the project as well as its image.

But for the complex to be economically viable, changes had to be made. The developers decided to use an agricultural greenhouse contractor rather than residential contractor for the atrium, which saved a lot of money.[25] The original goal of including rental social housing was abandoned, and the project addressed a moderate and upper-income group with units for sale rather than for rent, and the rental units served a moderate-income group as well. Also at time of sale the market for housing for elderly was momentarily saturated, so the housing was placed on the open market. But in the end, according to developer David Wolterbeek, "We sold all the housing. One guy 93 years old is still renting. People have the right to buy but are not forced to buy."

De Muzen park side showing curved roof

The developer was impressed by the architectural firm's high quality of design. "The architect and Atelier Pro had their designer rules and sustainability was included in the issues addressed." The project won a competition for sustainability for new dwellings in 1993, the only multi-family project selected.[26]

And a small, but important lesson learned in this project is about maintenance. The design of a building is tied to the structure and appearance of the building materials. If, for instance, the people maintaining a building do not attend to proper replacement of materials and the use of the original colors, a building will substantially change, and usually not for the better. In this case the darker stain used on the wood siding after the first few years (or possibly a change due to patina of time) substantially changed the appearance of the building, especially on the west side. More recently the original color has been used and the building looks almost like new. When a building is a landmark like this one, consistency with the original design is especially important.

The essential lesson here is that making an outstanding design come into reality and maintaining its quality requires commitment and teamwork on the part of the city, the architect, and for the long term, the developer, especially in the Netherlands, because the developer will own the building for the foreseeable future.

Notes

1. The information on Almere is taken from the Municipality of Almere website, http://english.almere.nl/the-city-of-almere/.
2. David Wolterbeek (Groene Almere/De Allantie, developer, De Muzen), Site Tour, February 23, 2015.
3. Noud te Riele (Atelier Pro, architect, De Muzen), Interview, The Hague, March 4, 2015, 4.
4. Ibid., 2.
5. Noud te Riele, Email Communication, July 13, 2016.
6. David Wolterbeek, Interview, 2015, 1.
7. Ibid., 1.
8. Noud te Riele, Interview, 2015, 4.
9. David Wolterbeek, Site Tour, 2015.
10. Ria van Dijk (De Muzen Homeowner and Urban Designer, Municipality of Almere), Site Tour, Almere, May 29, 2007.
11. Noud te Riele, Interview, 2015, 3.
12. Ria van Dijk, Site Tour, 2007.
13. Noud te Riele, Email Communication, March 4, 2016, 1.
14. Ibid., 3.
15. Noud te Riele, Interview, 2015, 10.
16. Ibid., 4.
17. Ibid., 2, 4.
18. David Wolterbeek, Interview, 2015, 3.
19. Noud te Riele, Interview, 2015, 5.
20. Ibid., 3.
21. David Wolterbeek, Interview, 2015, 2.
22. De Allantie, "Aangeboden door de Alliantie ALM—XML, Louis Davidsstraat 210, €135.000" (Real Estate Advertisement). WoningNet.NL Almere, www.woningnetalmere.nl/Eenheiddetails/150045946/3. ND
23. Noud te Riele, Interview, 2015, 3.
24. Jaap van der Laans, *Wonen in Almere 1.0: Almere als dagelijkse evaring* (Almere: Department of Urban Development (Dienst Stedelijke Ontwikkeling van de gemeente Almere), 2009), 75 [Translation by the author with assistance from Google.com].
25. David Wolterbeek, Interview, 2015, 2.
26. Noud te Riele, Interview, 2015, 3.

3 Vrijburcht

VRIJBURCHT

Location of Vrijburcht in the Netherlands

Location of Vrijburcht in Amsterdam

Location of Vrijburcht within Steigereiland

Vrijburcht is a very special project, having been developed as a cooperative housing project with prospective residents. Built in 2006–07 on Steigereiland in the IJmeer, an island specifically created for housing, this project comprises 52 dwellings, two guest units, a group home, four offices, a café, theater, childcare center, teen center, and community meeting space. The architect, Hein de Haan, worked closely with the homeowners' foundation, stichting Vrijburcht, to develop a site in Steigereiland that would serve not only the prospective residents, but also the broader community.[1]

De Haan developed an unusual approach to the design of housing in Amsterdam. After World War II, there was a severe housing shortage in the Netherlands. Beginning in the late 1960s and during the 1970s, young people in the cities who could not find affordable housing occupied or squatted in vacant industrial buildings. The government was sympathetic and passed laws to support the occupancy of abandoned structures under defined circumstances. By the 1980s, the Dutch economy had recovered from the war and cities in the Netherlands were able to reinvest in their neighborhoods. Since many of the squatted buildings did not meet code requirements, cities across the Netherlands together with private developers were interested in demolishing the old structures and creating new housing in their place. But in many cities, including Amsterdam, the squatters resisted and worked with some city officials, in this case aldermen, to transform the old buildings rather than tear them down.[2]

Much of the early part of Hein de Haan's architectural career was spent working with squatters, government officials, and housing corporations to transform industrial buildings into housing and workplaces for former squatters and others, especially artists, their families, and seniors. This work often involved developing new housing and workplace typologies to meet specific needs. But by the 1990s, most of the squatter

Northeast façade (front).
Photo by Joseph Messier

transformations were completed and de Haan, who took great pleasure in the challenge of working together with the future inhabitants of his projects, and in transforming existing neighborhoods, did not want to work with a regular developer in new residential neighborhoods as an architect who had no direct contact with prospective residents. Furthermore, he was interested in creating "functionally mixed areas with housing, working and public facilities."[3]

While the squatters were self-identified clients, now de Haan had to find clients to work with. So, de Haan began to design projects for people interested in working on cooperative housing. Thus began a new approach for de Haan that he later came to call "Collective Client Controlled Development" (a special form of what the Dutch call Collective and Private Patronage (CPO)).[4] From then on, de Haan with CASA Architects in Amsterdam designed numerous projects for client groups that he organized, first using the newspaper and later the Internet.[5] The group forms an organization (foundation, union, or cooperative) that elects a board to oversee the development process. In the case of Vrijburcht, de Haan and a group of artists responded to an announcement from the municipality of Amsterdam seeking the proposals for cooperative housing in Steigereiland. Their proposal was one of three chosen out of twenty-seven proposals.

The originating group began informally.

> We started at the kitchen table. That is rather special. . . . very significant for the atmosphere. . . . We started dreaming about living together with a number of friends—working and living together, sharing. We did it all together. . . . At that time it was hard to get a small workplace. And if you have a family it is very hard to divide your professional life and your family life. These were questions that could be addressed by living together. We wanted to propose a new way of living: the participation society. . . . [Hein de Haan made the first layouts from] the Lego blocks that belonged to his daughter. Every Lego block with a particular color represented the function of a space. And that is how he presented his first design to the local community, because there was no time. . . And that is the way we discussed our living. . . with colored blocks. And that is why I liked Hein so much. . . . he could make this project with Lego blocks.[6]

From the beginning, the Vrijburcht foundation was committed to creating a green building. In addition to treating the surface water for reuse on the site with a rain garden and storage tank, creating special wall pockets for nesting birds and having solar panels on the roof, the dwellings are equipped with in-floor heating and employ heat recovery in the ventilation.[7] Another initial commitment was to have a residence for disabled adults.

> One of the things I like best is De Roef, [the group home]. That is because the mother of . . . a girl with autism [had a] struggle to find a good place to live for her daughter. And it is almost impossible if you are critical about the circumstances that disabled people live in—you are not going to find a solution in the exiting institutions. I was more than pleased when Hein said "Do come with your ideas for a community for disabled people and we'll incorporate that in our project." And all the participants were very enthusiastic about that idea. . . . We can create more value, but another kind of value. For me that is the most important thing.[8]

One of the challenges faced by the initial organizing group, IJ-Burcht, was to be assigned a plot of land in the Steigereiland development. The city had designated a few lots to be developed by CPO projects, with the requirement that projects demonstrate organizational ability, financial viability, and contribution of something special to the neighborhood. The IJ-Burcht proposal for the biggest lot consisted of forty-two units as well as café-restaurant, theater, and childcare center, with Hein de Haan as an experienced organizer of CPO projects, was a standout in the group. Of the twenty-seven projects that submitted proposals for the designated lot in 2002, four were selected: Vrijburcht and three other proposals with about five units each. At this point, the groups faced a problem: the city had no system for dividing up the site. Eventually, the groups agreed to work together to create one complex under the leadership of de Haan, who was the most experienced developer.[9] Subsequently, two of the groups dropped out due to financial considerations, and the foundation, stichting Vrijburcht, along with the group Living Apart Together were created as the developing group.[10] Living Apart Together proposed to build five units but later dropped out, requiring stichting Vrijburcht and de Haan to quickly design ten replacement units.[11] This is an example of the flexibility required in such projects.

Another challenge faced by the group was to purchase rights to the land (Amsterdam owns the ground upon which structures are built and property owners pay an annual rent). The price was very high compared to projects that had recently been developed in the East Docklands, because the city had instituted a new system of calculating land prices, residual land value method, based upon actual costs to create the site and the value of housing to be sold. Because the developers in this area had planned high-quality housing, the land price was high, estimated at double the price of the land for the earlier projects. So, even though the city gave CPO projects a discount of 20% on the land, the price was steep, which contributed to the development fee, and the loss of a number of project participants.[12]

To reduce the cost of land, de Haan made a preliminary design that separated the workspace from the dwellings, allocating a good portion of the lot to a different address to be used as business space, which had a lower land purchase price than housing. This would allow a tax deduction, thus reducing costs for the businesses.[13]

The participants financed Vrijburcht themselves through stichting Vrijburcht. Investing approximately 8% of their housing costs, the members were able to "finance their development independent from banks or housing corporations," signing directly with the building contractor.[14] This fee covered all of the preconstruction costs. To share the financial risk, the elected board worked with De Key, a housing corporation that financed half of the apartments and agreed to serve as guarantor for unsold units. The housing corporation De Key owns and maintains the group home, café, and childcare center, which are managed by professionals; a volunteer foundation manages the theater; the Vrijburcht Homeowners Association, as required by law, manages the general shared project spaces; and the Vrijburcht Foundation manages the special areas such as the guest rooms, garage, hobby space, and docks.[15] In the end the housing cost was 10–18% lower than comparable units in IJburg.[16]

Site, Context, and Building Functions

As a part of the IJburg Haveneiland development, Steigereiland is an island created from sand from the IJsselmeer and the IJ just east of Amsterdam. Haveneiland, designed as an extension of the city, is a

Greenhouse and teen center

Childcare center

Theater

Café

long, thin series of island with grids of canals that allow every dwelling to be in close proximity to the water and to a central tramline that connects directly to the central station in Amsterdam in fifteen minutes.[17] The streets are organized in a Manhattan grid, with commerce along the main artery, public space consisting of "[waterway] banks, streets, canals, bridges, parks and locks," and three public gardens and a park.[18] As with most developments in the Netherlands each type of movement has its own path, whether pedestrian, automobile, bike, or tram. With a floor area ratio of approximately 1, and an overall density of 70 units per hectare (28 units per acre), Haveneiland is double or triple the density of a typical suburban Dutch area, although lower than that of the East Docklands and historical areas of Amsterdam.[19] Being built on what are human-created sandbars, the island sites are essentially flat.

Vrijburcht is similar to other housing in IJburg, having been developed as a large block, although it has a much greater diversity of housing than the typical block, and slightly higher density. For instance, each of the adjacent Steigereiland blocks is composed uniquely of rowhouses, apartments or freestanding houses, while the Vrijburcht block is composed of a variety of housing types.

The planning group's vision of Vrijburcht was as a complex that would be a hub for the surrounding neighborhood as well as for its own residents. This drove the inclusion of the childcare facility for forty-five children, the café, the theater, and other community spaces such as a swimming dock and a dock for small boats.[20] Additionally, with its live-work units and on-site employment, Janssen and Jansen report that approximately forty-five people work at the site for about forty full-time equivalent positions.[21] They note that the type of work is largely creative and service-oriented, such as design, advertisement, business

Vrijburcht:
a. location of community spaces
b. horizontal circulation, and
c. vertical circulation

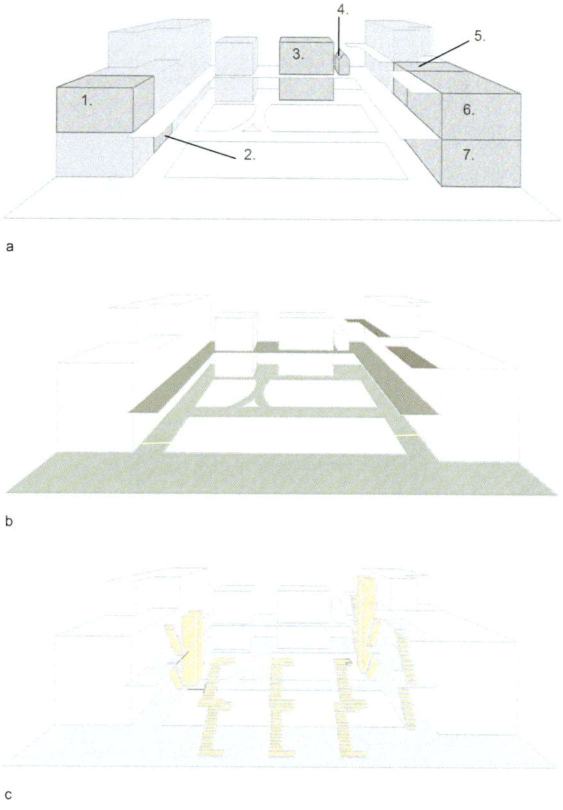

administration, accounting, massage, and psychotherapy.

Although the original plan was for a relatively small project, the process of site selection generated a much larger number of units which in the eyes of the residents was fortuitous, as 52 turned out to be a kind of magic number that was large enough to develop certain economies of scale while still having a scale where all the residents could know each other.[22]

> It is a very nice group, about 50 apartments. . . . [At] between 80 and more than 35, we have the magic number. At 50 you're small enough to know each other and big enough to hide away.[23]

Organization, Appearance, Form, and Massing

As described earlier, Vrijburcht comprises an urban block with housing on four sides that enclose the communal courtyard. Two opposite sides of the project, southeast and northwest, with the exception of a few flats, are designed as two layers of maisonettes, with live-work units at ground level. In between, the northeast side along the main street includes live-work units, the childcare facility, and community spaces above, with the café and the community center or theater on the south corner. The southwest side on the IJmeer is made up of water-facing flats with balconies toward the sunset. Automobile parking is provided underground, with access from the street at the northeast corner; and bicycle parking is on the northeast side at the courtyard level between the childcare center and the atelier units. Additionally on the IJmeer are the boat dock and boat storage area as well as a swimming dock.

The courtyard level of the complex serves as a kind of datum for socializing, as this is where the elevator from the parking connects to the east and west elevators (as well as a number of staircases). The elevator on the west connects the courtyard to the first level northwest deck, and the east elevator connects to two levels of circulation deck, the first and second. The organizational principle is that workspaces are on the ground floor accessible to the street, with living spaces above. The only dwellings at the ground level are the water-facing lightwell units on the southwest side that arrive at their front door via three stairwells,

Vrijburcht floor plans

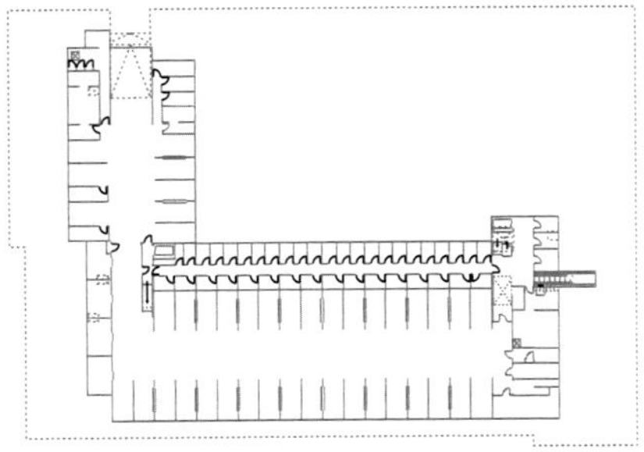

Parking Level

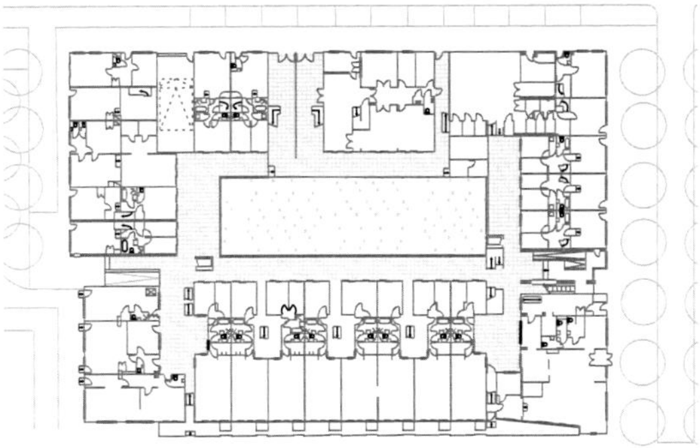

Ground Level

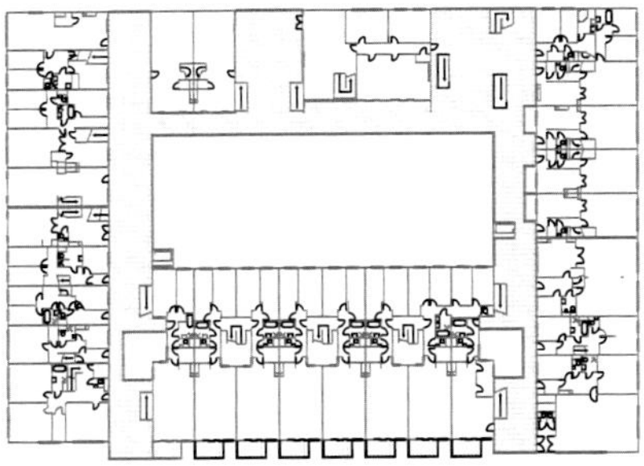

Level 1

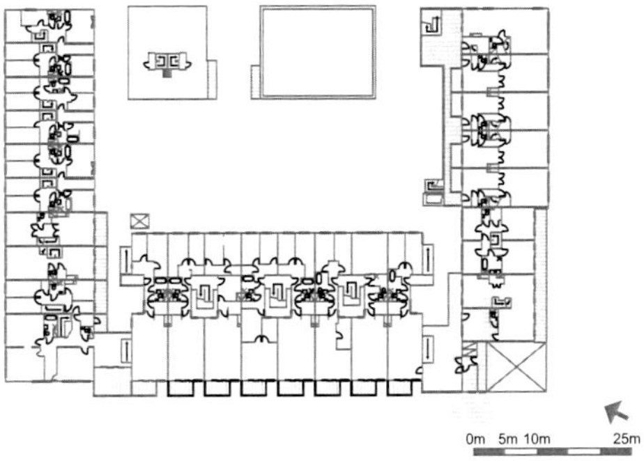

1:1500

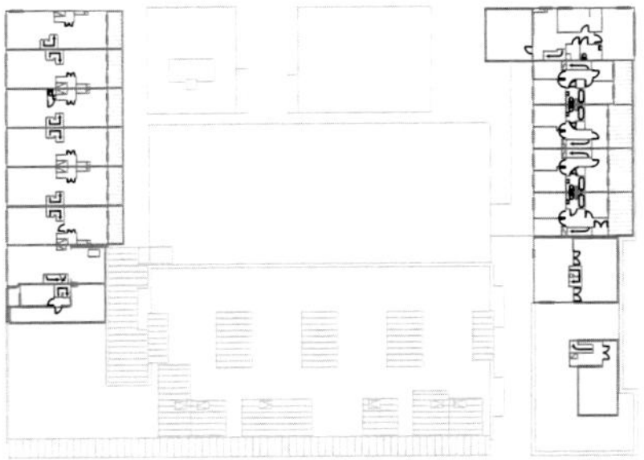

Level 3

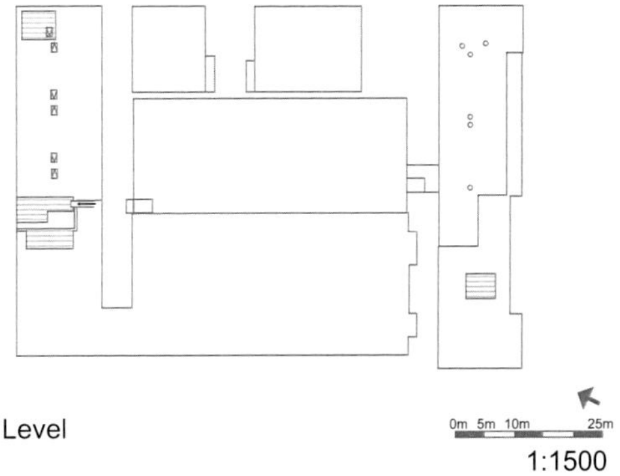

Roof Level

1:1500

Lightwell access stairs

Extra-wide northeast level 1 deck

and a few units on the southeast corner that are not connected to the street.[24]

At the first level above the courtyard, the southeast exterior deck is both circulation and social space, with seating along its southeast side that overlooks the courtyard. The northwest first level deck is more private, with entrances to maisonette units mixed with the outside space for some of the units. The extra-wide decks on the first level above the ground allow both entry and terrace to co-exist. There is sufficient space for tables and/or chairs next to the walkway. At the second level on the northwest the narrower access deck connects to the lower level entries of the upper maisonette units that have balconies.

The three sides of Vrijburcht that face streets form a rather impressive red brick wall that resembles a fortress like its name, Free Citadel or Free Castle. While parts of the southeast and northwest sides of the project are four stories tall, most of the front northeast façade is two and three stories and the rest of the project is three stories in height. But within the walls, from each of the surrounding streets on the southeast, northwest, and northeast are three large entries that give an inviting peek at the green courtyard from the exterior.

The courtyard at Vrijburcht is perhaps the most communal of all the courtyards in the study, since it is furnished with furniture for gathering with moveable picnic tables, includes a green space with play area and rain garden, and is visually accessible to the childcare center and other amenities as well as all of the dwellings in the project (see photo on page 74). The courtyard is also activated by its many entrances and overlooking access decks. On the ground level, the walkway through the courtyard connects all seven segments of the project. Thus the social life of the project focuses on the courtyard.[25]

Southeast level 1 deck, looking southwest. Photo by Joseph Messier

Northeast and northwest façades

The size of the courtyard at Vrijburcht is relatively small and intimate compared to others. It is sunny as well, due to the low surrounding buildings—on the southwest side only three stories tall. Because the underground parking is beneath the housing rather than the courtyard, the trees in the courtyard have ample room for their roots. The courtyard also serves as a rain garden with a tank beneath, where the water is treated for reuse as irrigation.[26] One day that I visited, children played in the rain garden complete

Panorama of the courtyard northwest side. Photo by Cynthia Lapp

with rubber boots and a net to catch frogs (see photo on page 80).

The childcare center or crèche is on the ground level located along the main, northeast segment of the building, and opens southwest into the courtyard where the children play. Above the crèche are the community and teen space in a greenhouse that is reached from the level 1 deck. Located in the south corner, the theater and café serve the entire Steigereiland community. The café on the ground level has its own terrace for outdoor dining, with the theater above it entering from the first level deck. Two guest units on the first level deck enter from the deck as well as from the theater, serving as backstage.

Housing and Unit Typology

Vrijburcht intentionally has a wide range of housing types including flats, live-work units, maisonettes, and also a group home. The resident design group sought inhabitants with a variety of lifestyles and ages, including their young adult offspring with mental disabilities. Although the design participants themselves could finance their own units, they sought to include others who would purchase housing and live-work units on the market as well as those who would need to purchase with subsidies. Thus, they contracted with De Key, a housing corporation that develops and maintains social housing, to sponsor the group home. Stichting Vrijburcht sponsored and developed the ten affordable units as AMH housing (Amsterdam Middle Segment Mortgage housing for people with incomes between \$20,000 and \$40,000 euros).[27] In the end, all 52 of the units were sold to people with a wide range of incomes."[28]

Hein de Haan worked with each household to create a unit to fit their needs.

> Hein made a kind of model, a very simple one. And he discussed the apartment with everyone who wanted to have one, [asking] "What do you want?" and was changing the walls. Nearly every apartment is a little different from another one because of those discussions.[29]

Vrijburcht unit plans: A

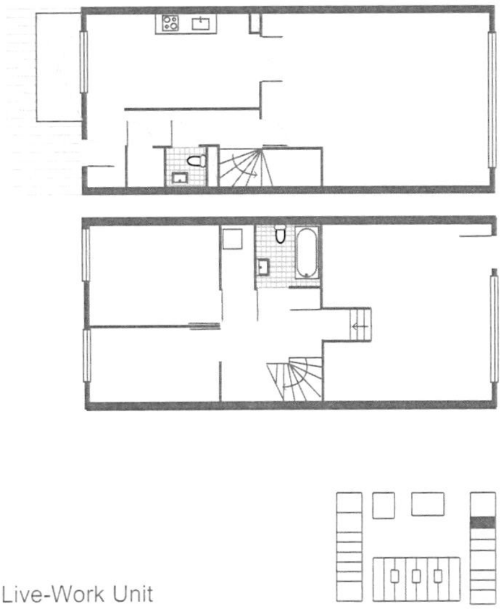

a. Live-Work Unit

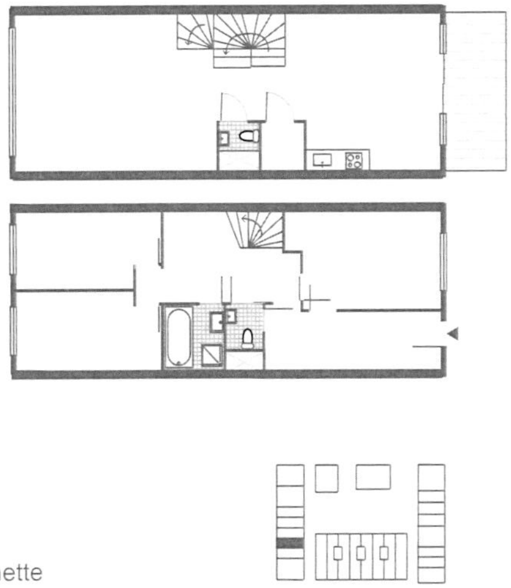

b. Maisonette

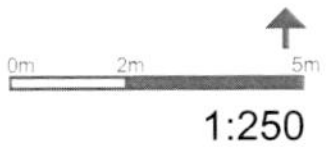

The units on the ground floor along the streets on the southeast, northeast, and northwest are largely designed as workspaces that link to dwellings. One of these is a flat with a second access from the courtyard level, and several others are workspaces without a connected dwelling, but the rest are maisonettes that have work access from the street and residential access from the first level deck. Above these live-work units are flats and maisonettes that have their access from the first level deck. On the southeast side an additional level of maisonettes has its own access deck at the second level above the ground.

Although the wing on the northwest side is also four stories in height, unlike the units on the southeast, all of the maisonettes are reached from the first level deck, requiring some unit residents to walk up two flights of stairs to reach their living room, and three to reach their bedrooms. However, these units are compensated by having both private decks overlooking the courtyard and access to the roof above. Owners of four of these dwellings have built decks on the roof.

On the southwest side of the complex are the twenty-three lightwell flats in three layers that overlook the water of the IJmeer.

Live-Work Unit

The ten live-work atelier units at Vrijburcht are essentially maisonettes with workspace and work entry at the street level, a half-level below the courtyard. Thus, the workspace is clearly delineated from the living area (see top photo on page 76). The living space, consisting of living room, kitchen and toilet is entered from the level 1 deck (see bottom photo on page 76). At the courtyard level are the bathroom and bedrooms, either two or three, depending on the design. Five such units are on the southeast side, two on the northeast and three on the northwest. The dwellings' outdoor spaces are on the wide entry deck at the unit entrances.

Atelier leading to bedroom corridor (live-work maisonette)

Living room of the live-work maisonette

Several of these units have a unique design. For example, an atelier at the northeast corner is three levels, with only workspace at the street level, two bedrooms and secondary entry at deck one and living spaces and formal entry at level 2. The adjacent unit does not have bedrooms on the courtyard level, but has two additional bedrooms at the living area level. In addition to the ten live-work atelier maisonette units, Vrijburcht incorporates four work-only units and one live-work flat. The two work units and the live-work flat located at the west corner of the project are all on one level, being entered from the street by means of a short stair.

Maisonette

Eighteen of the units in Vrijburcht are two-story maisonette dwellings that enter either directly from a circulation deck or indirectly from a staircase that is entered from a deck; most of these stairs are internal, but a few are outside (see plan b on page 75). These maisonettes are on either the southeast or northwest wing of the project. Seven of the eight maisonette residences in the northwest block have the bedrooms on the lower level and the living areas above. Their balconies open from the living area, and in three cases they have a roof terrace. The eighth dwelling is a three-story unit with bedrooms at deck level 1, living areas at level 2, and extra room and roof terrace on level 3.

The maisonettes on the southeast side are mostly organized with living spaces on the lower level and bedrooms above. Four of these enter from the level 2 deck overlooking the courtyard, and at the level above have three bedrooms and balconies facing southeast. A fifth unit, at the east corner, enters from the level 2 deck, but has all of its living and bedroom space at level 3, as well as a large terrace. This unit, Dutch-style, has two rooms in addition to the kitchen-dining area that could either be arranged as two bedrooms or as one bedroom and a living room.

The two west wing maisonettes closest to the southeast end enter directly from the deck at level 1. One of these combines what were originally two units. The extra-large living room area at level 1 connects to two independent pairs of bedrooms on level 2 and to a shared living space on level 3 that opens onto a large southeast-facing terrace (see photo on page 77). The dwelling located adjacent to and over the theater has bedrooms at level 1, but enters from a special deck and staircase at level 2. This unit has a long balcony along its street

Terrace for the double-wide maisonette

Vrijburcht unit plans: B

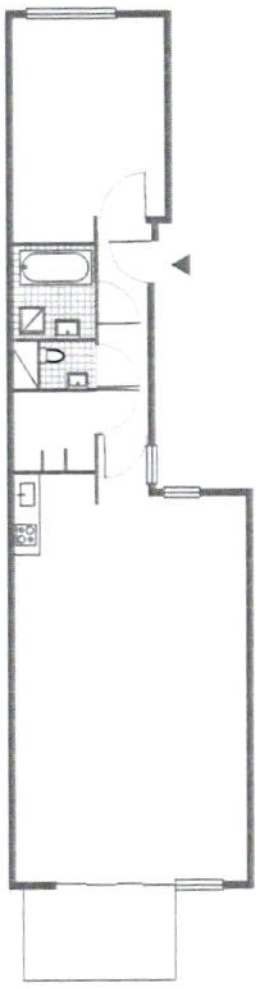

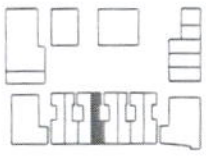

c. Lightwell Flat

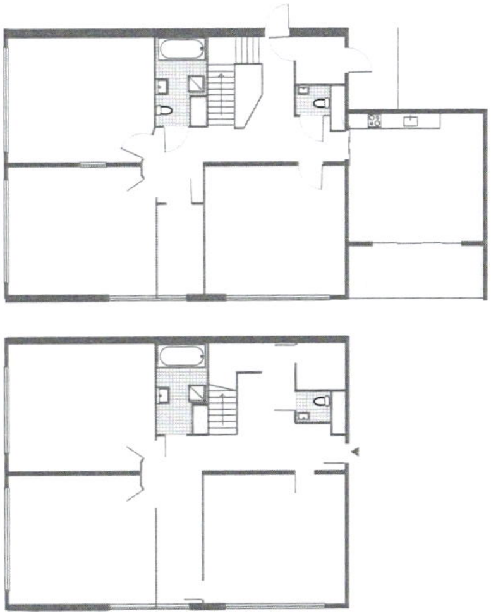

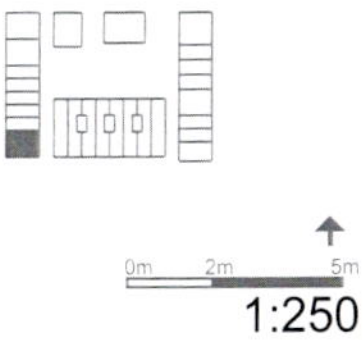

d. Group Home

façade at the living area on level 2 and access to a roof deck on level 3.

Lightwell Flat

All twenty-three of the flats that are not live-work are located along the southwest side of the building facing the water (see photo page 78). In an area eight bays wide and three levels high, all of these apartments have southwest facing balconies overlooking the water (see plan c to the right), and, with two exceptions (the three-bedroom and efficiency units), bedrooms with windows on the courtyard or northeast side. Most have two bedrooms, but three have four bedrooms, one has three bedrooms, six have one bedroom, and one has no bedroom; the bedrooms themselves vary in size.

Most of these dwellings are reached from the three stairwells located in the light-wells and originating at the courtyard level (see photo page 72), although the units on the southeast and northwest ends enter from courtyard and deck levels, either directly or by means of their own staircases. While most of the flats require stairs to access, eleven are handicapped accessible due to being on the ground level or directly accessible from a deck. As an example of the variety of lightwell dwellings, the three

Lightwell unit balconies face the water

accessible flats at level 1 include one two-bedroom, one four-bedroom, and an efficiency unit.

At the courtyard level, due to the need to gain access to the stairs in the lightwell, half the dwellings have one bedroom and half have two. With the exception of the four-bedroom and efficiency units, all of the flats on level 1 have two bedrooms.

The level 2 dwellings have the most variation, with the two flats that have four bedrooms adjacent to the only three-bedroom unit, that also being the only unit with a bedroom facing the river. In addition to the southwest balcony in each unit, all but one of the seven units at level 4 has a roof deck.

Group Home

The group home, De Roef, at the west corner of the complex, is on the first and second levels of the project, entering on level 1 (see plan d on page 77). Designed to house six young people with mental disabilities and a staff person, it provides each resident with his/her own room, front door, and doorbell. The community area comprises a kitchen, living area, and shared bathing facilities. An adjacent flat unit houses staff.[30]

Guest Units

Since many of the residences are small, the original planning team members decided to include two guest rooms that can be leased. Adjacent to and connected to the theater, these rooms share a WC, bathroom, and kitchen, thus allowing them to serve both as dressing rooms and kitchen for theater events, and also to house visiting artists associated with the theater. Residents may also rent these for their guests.

Syntactical Structure

Despite being one of the smallest projects in this book, Vrijburcht exhibits a highly complex and connected syntactical structure. Although much smaller than La Grande Cour, it is similarly complex. Its level of connectedness resembles De Zilvervloot. The complexity is due to the variety of circulation spaces in the project (dark dots on the diagram) that represent 27% of the circles on the diagram. While many units enter directly from the street, courtyard, deck one, deck two, and the three decks on level 2, other entrances are reached indirectly through stairs from the courtyard or deck levels.

There are several unusual patterns in the diagram, one of which is created by the ten highly connected atelier units that enter at both the street and deck one at the

Vrijburcht space syntax gamma diagram

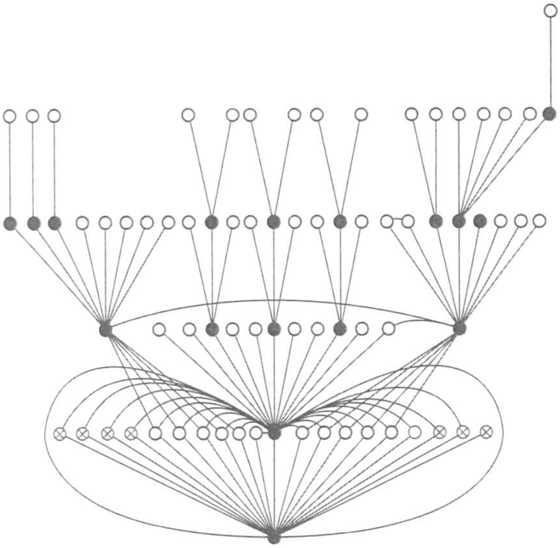

lower left side of the diagram. At the upper left side of the diagram are several threads created by dwellings that are reached by means of stairs from deck level 1, as well as three threads of units that enter from three different level 2 decks. Three of these threads at the right side of the diagram represent the lightwell units that are connected through the three sets of stairs. A third pattern is the set of three threads that represent access from the three separate decks at level 2. These connect to one, four, and eight dwellings.

The courtyard provides a central place for all residents to gather. Because almost everyone has to traverse it to get to their home, and all the dwellings are connected to it through the unified circulation system, the courtyard location encourages residents to recognize each other and to socialize. This raises the question of density. The arrangement of the units in this project at density of 118 units per hectare (49 units per acre) effectively supports a highly interactive community. None of the other projects is designed to sustain this level of interaction. It may be difficult to sustain high community interaction if the number of units per hectare is increased.

The connections are also unusual, since twelve of the units and the theater connect to two different levels, and all of the circulation paths connect to each other. None of the other projects in this book has this kind of within-unit connection to two floors, although De Opgang, De Zilvervloot and La Grande Cour have connections between circulation systems. Here the courtyard is the primary connection (black dot on level one in the middle). Not only does it provide access directly to the eight lightwell units on the ground floor, and indirectly to the twelve other lightwell units that enter from the three stairs, but also the courtyard must be traversed to reach all of the other units. The second most important connection points are deck one and two, which connect directly or indirectly to the remaining twenty-nine units (see the black dots on the far right and far left of level three that connect both up and down).

Social Principles

Vrijburcht has a great diversity of housing types and other spaces because of the commitment of the founding group to having a cooperative with a varied age group, income level, and a range of lifestyles and abilities, as well as work and community activities. The housing encompasses smaller units for singles and larger units for families. The work studios and live-work ateliers provide opportunities for small entrepreneurs like graphic artists to be a part of the complex. The complex provides affordable housing, middle-income housing, and even a few luxury dwellings as well as a group home. Vrijburcht supports families by providing childcare, courtyard play space, workshop, and extra facilities such as the theater, docks, and sailing school. Its accessible units and group home support people with varying abilities (although the number of dwellings with entrances requiring stairs is high at thirty-eight, all community spaces are accessible).

Child playing in the courtyard rain garden. Photo by Joseph Messier

The foundation supported two art pieces by the sculptor Suzanne Willems, one called "Bootje" (baby boat) on the quay, and "Fishes" on the southwest façade (see the right side of photo on page 81).[31,32] Finally, by providing community spaces, Vrijburcht serves as a locus for the wider neighborhood.

The entrances to units are structured throughout the project in small groupings of dwellings, such as the six lightwell units that collect at each of three staircases, the ten units and six group home residents that enter directly or indirectly on the northwest side of deck level 1, the ten maisonette and lightwell units that enter directly from the northeast side of deck level 1, and the four units that enter directly from deck level 2 on the southwest. This not only encourages social engagement, but also provides security, as residents will immediately recognize an outsider.

Another social principle exemplified in Vrijburcht is environmental concern, as represented in the water collection, heating systems, and nest boxes for birds designed into the façade.[33] Additionally, Vrijburcht is embedded in the community because the childcare, theater, café, boat, and swimming docks serve the entire neighborhood, not just the complex.

Lessons and Conclusions

The most important lesson of Vrijburcht is that a group of people working with an architect can create a collective housing project that encompasses a wide variety of people of different income levels and ability into one project, and also that incorporates enduring community activities like the theater and community service like the café and childcare center. The success of Vrijburcht as a social enterprise is undeniable. Not only does its highly connected internal design nurture the residents, but it activates the street with workplaces, and serves the neighborhood with the group home for disabled people, the childcare center, café, theater, dock, and sailing school. These extra programmatic elements are what sets CPO development apart from other projects. As explained by Hein de Haan:

> The great advantage of CPO is the programme it has to offer to the rest of the city. . . . A range of mixed urban activities has been accommodated in this block, and so it means something to the neighborhood. A block like this really adds something to the city.[34]
>
> With CPO, you not only can introduce a richer programme in the city, but you can also offer accommodation to income groups and forms of living that normal, regular housing developers do not take into consideration at all. Take the residential/work combination: a simple owner-occupied property with two front doors so that the home and the office each have a separate entrance and the office can be registered at its own address.[35]

Vrijburcht shows that it is not only possible to develop a successful cooperative community, but that in the Netherlands it is possible to do so using future residents' capital investment with the cooperation of a housing corporation, and to do so at a final cost of housing that is lower than housing achieved through the normal development process. That achievement however, was

Vrijburcht southwest façade

not inevitable. The foundation for the project's success lies in the investment of time and expertise provided by the organizing group and in particular the accumulated experience about such projects of the architect Hein de Haan.

In the Netherlands, there is a series of institutional structures designed to enable cooperative housing to be built. First in this particular case is the city of Amsterdam, which was interested in supporting collective housing in the new urban development, and offered the Steigereiland site. Because Amsterdam owns its land, the municipality can control how the land is leased, at what cost, and to whom (this is not the situation for most cities in the Netherlands). A second institution that can assist is the housing corporation. In this case, De Key was willing to step in early in the process to reduce the financial risk. A third institution was the Amsterdam Middle Segment Mortgage (AMH), the regional program that formerly supported people with moderate means to purchase a dwelling. This instrument provided the financing for stichting Vrijburcht to develop the low- to moderate-income housing in the project.

These structures however, did not perform perfectly. As discussed earlier, during the process of assigning land leasing rights on Steigereiland there were some difficulties in determining which cooperative owned how much land. Also, during the development process, when it came time for AMH buyers to purchase their units, there were administrative barriers. The city of Amsterdam had a policy of not endorsing AMH loans and the national program required municipal endorsement. Eventually, in partnership with Rabobank Amsterdam, the paperwork was completed and Vrijburcht sold all of the dwellings step by step. The difficulty of finding investors willing to finance cooperative housing is a major challenge to the development of such projects. Furthermore, at several points in the development process when additional investment was required, some participants had to drop out.

In regard to the final, lower cost of the units, there were several contributing factors. First was the form of the contract with the builder. Normally the housing corporation would take the contracted working drawings and get competitive bids from builders. In this case, however, to speed up the process, the foundation stichting Vrijburcht created a multiple negotiated contract prior to the development of working drawings. Four contractors were invited to make a fixed-price offer based on the drawings of the final design and a technical description. Stichting Vrijburcht then worked together with lowest bidder, BK Bouw, to create the working drawings that formed the basis of a fixed-price contract prior to beginning construction that was very economical. Several other factors contributed to the reasonable cost of the dwellings, among these the density of the housing on the site, the courtyard arrangement that reduced the use of elevators to a minimum, the basic structural system materials chosen, and specifically the use of wood decks rather than concrete for the access (a reduction of 66%).[36]

One factor that did not contribute to the lower cost of the dwellings was the land cost, as discussed earlier, which despite the special purchase rate given to collective housing was much higher than previous developments due to the newly instituted residual land value method for estimating cost. In the end, the very high cost was mitigated by the fact that not all the property was devoted to housing. The civil servants ended up evaluating the land using the description that had been provided by the foundation that identified four different uses for the space: market-rate housing (calculated as housing), working space (calculated as commercial space), community space (donated by the city), and the AMH housing units (calculated at a reduced ground price to make them affordable).[37]

As can be seen from this discussion, the success of the project can be attributed to the time and energy devoted to addressing all of the details. But the success can also be credited to the imagination and savvy of the architect, Hein de Haan, who conceived of a project that would appeal to the land-assigning jury by including a large number of dwellings as well as on-site workplaces and community amenities. As described by Janssen and Jansen, the Vrijburcht proposal "jumped out" in competition with the other collective projects being an "ambitieus, groot plan dat diversiteit brengt in het programma met meerwaarde voor de buurt via activiteiten in voorzieningen en horeca" ["ambitious master plan that brings diversity to the program with added value for the neighborhood through service facilities and hospitality"].[38] Moreover, due to de Haan's involvement, experience and proven record, the municipality was happy to support what promised and proved to be a successful project.[39]

In sum, while it was a struggle to get the project built, the outcome met the goals of the inhabitants, almost all of whom are still living there after nine years at the time of this writing.[40]

Notes

1. Rainer Janssen and Nico Jansen, "Vrijburcht: Amsterdam—Oost," *Vitale Knoppunten van Zachte Infrastructure*, www.knooppuntenvanzachteinfrastructuur.nl/portfolio/vrijburcht/ (Berlin and Amsterdam, 2013), 3.
2. Hein de Haan (CASA, architect, Vrijburcht), Discussion, May 30, 2007.
3. Hein de Haan, "Collective client controlled development of space: Examples from an Amsterdam practice," in *Making room for people: Choice, voice and livability in residential places*, edited by Lei Wu and Evert Hasselaar (Amsterdam: Techne Press, 2011), 130.
4. Ibid., 130.
5. Hein de Haan, Discussion, 2007.
6. Annelie Seegers (resident, Vrijburcht), Interview, Amsterdam, June 16, 2016, 2.
7. Jansen and Jansen, "Vrijburcht: Amsterdam—Oost," 9.
8. Annelie Seegers, Interview, 2016, 2–3.
9. Ibid., 3.
10. Menno Vergunst and Johan Vlug, "Vrijburcht," in *CoHousing cultures: Handbook for self-organized, community-oriented and sustainable housing* (Berlin: Jovis Verlag, 2012), 169.
11. Johan Vlug, Personal Communication, May 1, 2016.
12. Janssen and Jansen, "Vrijburcht: Amsterdam—Oost", 5.
13. Ibid., 3.
14. Vergunst and Vlug, "Vrijburcht," 134.
15. Ibid., 169–171.
16. Janssen and Jansen, "Vrijburcht: Amsterdam—Oost", 7.
17. Vergunst and Vlug, "Vrijburcht," 167.
18. Ton Schaap, *Amsterdam* (Rotterdam: 010, 2008), 26.
19. Ibid., 16.
20. Hein de Haan, Discussion, 2007.
21. Janssen and Jansen, "Vrijburcht: Amsterdam—Oost", 13.
22. Annelie Seegers, Interview, 2016, 3, Johan Vlug (resident, Vrijburcht), Interview, Amsterdam, June 10–11, 2016, 2.
23. Ibid., 3.
24. Johan Vlug, Personal Communication, 2016.
25. Vergunst and Vlug, "Vrijburcht," 171.
26. Vergunst and Vlug, "Vrijburcht," 165; Hein de Haan, Discussion, 2007.
27. Anja van Heelsum, *Case study on housing: Amsterdam, Netherlands* (Dublin: European Foundation for the Improvement of Living and Working Conditions, 2007), 20.
28. Janssen and Jansen, "Vrijburcht: Amsterdam—Oost," 7.
29. Johan Vlug, Interview 2016, 5.
30. Woondroom Zorg, "De Roef," Utrecht: Woondroom, 2014, woondroomzorg.nl/samenwerkingspartners/deroef/tabid/397/default.aspx.
31. VLUPG, "Vrijburcht: Collective and private patronage—also for public space," *VLUGP Stedebouw & Lanschaparchitectuur* (Amsterdam, 2010), www.vlugp.nl/wp-, 5.
32. Johan Vlug, Personal Communication, 2016.
33. Vergunst and Vlug, "Vrijburcht," 165.
34. Hein de Haan in Joosje van Geest, "A generation that is growing up that can't even share a single facility: Interview with Hein de Haan, CASA," *Dash #8*, March 1, 2013, 49.
35. Ibid., 51.
36. CASA architects, "Answers to Questions," Written Communication, March 14, 2016, 2.
37. Ibid., 1.
38. Janssen and Jansen, "Vrijburcht: Amsterdam—Oost", 12.
39. Ibid., 12.
40. According to founding residents Annelie Seegers and Johan Vlug, one owner sold the unit immediately, and a few other moves have been the result of family situations (such as divorces) accounting for about five of the fifty-plus units. (Annelie Seegers, Interview, 2016; Johan Vlug, Interview, 2016).

4 De Zilvervloot

Polski sklep
supermarkt
Dagelijks vers & kwaliteit

De Zilvervloot east façade: Het Want

A medium-high density project with 120 units per hectare (40 units per acre), De Zilvervloot is located in Wielwijk, a neighborhood of the city of Dordrecht. De Zilvervloot is notable because it was developed to regenerate a neighborhood, and architecturally is one of the few projects in this study that intentionally expresses complexity in its design. Its design celebrates diversity in terms of the design of units, of its architectural expression with many colors and materials, and even in the mix of types of residents, to the point that residents of luxury flats may share elevators with residents of social housing units. Here the combination of residential and commercial uses within the complex generates a lively commercial center for the formerly exclusively residential area.

The Wielwijk neighborhood was built in the 1960s by the housing corporation Woondrecht as social housing, a practice typical of the time. By the 1980s the exclusively low-income quarter had become dangerous and drug-infested. Realizing a need to radically change the neighborhood's image, Rob Hagens, then a director of Woondrecht and responsible for the site, convinced the housing corporation of the need to transform the neighborhood, and the value of involving the residents in the change.[1] As a result of a competitive hiring process, Atelier d'Urbanism, d"Architecture et d'Informatique (AUAI), a firm well-known for participatory community design, with architects Lucien Kroll and Dag Boutsen, was hired to redesign the neighborhood and the buildings.

Responding to several years of meeting with the community residents' team, the AUAI architects created a plan that incorporated new housing for a range of incomes, including housing ownership for upper-income households. The neighborhood design encompassed renovation of a housing project for the elderly, transformation of an existing social housing project into a mixed-income project, adding new sports facilities, adding a school, adding grocery stores and

Location of De Zilvervloot in the Netherlands

Location of De Zilvervloot in Dordrecht

Location of De Zilvervloot within its neighborhood

East façade facing the Admiraalsplein

commercial space, and creating the entirely new mixed-use, mixed-income housing block called De Zilvervloot. The De Zilvervloot complex, completed in 2005 as a pivotal project in the neighborhood redesign process, includes about 50% social housing for rent, with the 25% luxury units and 25% units for middle-income residents, being a mix of rental and units for purchase. The site is the former Admiraalsplein, a very large open area that had been unused and too large to be easily supervised. Instead, a smaller, human-scale Admiraalsplein to the east of the new Zilvervloot development serves as an urban plaza for the neighborhood.[2]

De Zilvervloot's 130 dwellings (105 apartments, eighteen maisonettes, three penthouses, one studio, and three rowhouse dwellings) are located in nine buildings built on top of a 150-auto parking garage and a layer of ground-level commercial spaces ranging from supermarkets to restaurants. The form of the building, with its two courtyards, is a result of discussion and compromise between the architect and the community participants.[3] This urban housing block achieves a density of 170 units per hectare (69 units per acre).

Design Approach

Not only did the AUAI design methods involve citizen participation, but the architects also used the Open Bouwen or Open Building approach developed in the

De Steven tower on the southwest corner

Netherlands.[4] Additionally, the architects were concerned with applying the most advanced ecological principles for renewable and healthy energy-conserving materials, with energy production on site in the form of solar panels, and with green roofs that also serve as gardens for the units on the residential courtyard and upper levels.

The Open Building approach is ecological (in the sense that it accommodates change with less investment of materials and energy) and human-centered (facilitating adjustment over time to changing building occupancy). Premised on the observation that different aspects of our built environment last for different lengths of time, Open Building allows for different levels of permanence and change in built settings. Whereas streets often endure for millennia (consider the Roman roads), furniture within rooms is likely to be adjusted almost daily, and the spaces within dwellings to be changed every 10–20 years. Applying an Open Building approach to the design of multi-unit housing, the architect designs a building's exterior, shared circulation, heating and plumbing infrastructure to serve a long time, while the interior layout of the unit can be easily modified without significant expense.

Site, Context, and Building Functions

The 0.77-hectare (1.9-acre) De Zilvervloot site lies at the edge of the city of Dordrecht on a main street and major bus line. Its location at the former Admiraalsplein is the heart of the 131-hectare (324-acre) neighborhood of Wielwijk. The topography of the area, typically Dutch, is essentially flat.

The construction of De Zilvervloot was an initial phase of a major effort of the city and the district to rejuvenate the entire Wielwijk neighborhood, and give it a new identity within Dordrecht. The larger plan involves such ambitious changes as diverting pass-by traffic from the main street to a nearby highway, creating more visible water systems, improving access to parks by creating a green path through the district, increasing and greening bicycle access, and adding new sports areas, as well as renovating and reconstructing existing residential buildings.

De Serre galleries, De Steven tower, and the residential courtyard

addresses the community's interest in having a neighborhood made up of people with a variety of lifestyles and incomes. Together De Zilvervloot and the Admiraalsplein form the heart of the Wielwijk neighborhood.[5] Surrounded by newly built and renovated mixed-income housing on all sides, the complex and plaza and the adjacent new sports facilities and community space thereby serve as the neighborhood's commercial and civic hub.

The special fit of De Zilvervloot with the area around it is a result of twelve years of architects and the community working together.[6] One response to stakeholder's concerns took place after the community reached agreement on having a large new housing block in the new neighborhood with several courtyards and a tower. Placement of the housing tower on the site of the new De Zilvervloot complex challenged the group. After much discussion and examination of models and renderings exploring mass and daylight, the group, wishing to support the neighbors, decided that the most important criterion was to have no shadow on the existing adjacent apartment buildings. As a result the tower's moving shadow hits only the courtyards and housing within the project and not the next block. The tower also marks the landmark status of this project within the neighborhood, while the new commercial courtyard invites all residents to engage with the new building.[7]

The inclusion of commercial space at the lowest level of De Zilvervloot was an essential part of the neighborhood transformation, as was including sports facilities and community space in adjacent structures. De Zilvervloot with its new shopping area takes up the western half of the former unused open area, with the remainder reformed as the human-scale plaza with water feature still called Admiraalsplein. The nine buildings of De Zilvervloot's housing enclose both a residential and a commercial courtyard. The housing

Organization, Appearance, Form, and Massing

In contrast to the unified façades of most other housing complex housing projects shown in this book, the exterior of the housing in De Zilvervloot is partitioned into nine buildings, each with its own material, fenestration, and color palette. Furthermore, the buildings are richly articulated with balconies, roof gardens,

De Zilvervloot floor plans: parking to level 2

setbacks, and material changes, while the individualized massing of the housing creates an irregular roofline, resulting in a very high level of visual intricacy compared to other projects. In fact, Lucien Kroll takes an incrementalist approach rather than rationalist, and his buildings express a certain lack of predictability. "If you want to plan something reasonable, you get into an adventure. It becomes impossible to manage. This creates complexity."[8]

The nine residential buildings, located over underground parking and over the grocery stores to the west, frame the two courtyards. At the east of the site, adjacent to the plaza and joined to it through a housing bridge, lies the commercial courtyard. Two grocery stores and a restaurant are accessible through this courtyard while the remaining smaller retailers are located along the street to the north. The western residential courtyard, surrounded by housing on all sides, is inaccessible to the public but

De Zilvervloot floor plans: levels 3 to 6

accessible to all of the units in the complex through interconnected corridors.

Access to light from more than one side is one ordering principle for the buildings. Most dwelling units penetrate the building section to have light on both ends, but others are on a corner with light from two sides or on an end with light from three sides. Another principle is that every unit has at least one outdoor space, whether a balcony, loggia, or roof terrace. Because the architect is philosophically opposed to having identical units when each family is unique, the building is arranged so that opportunities for variation, whether in the design of the front door, the shape of the rooms, or shape, location, and type of outdoor space, are maximized.[9] Nonetheless, the project came in within budget, although some of the original material choices had to be changed, such as brick being replaced by stucco and metal panels.[10] The

De Zilvervloot floor plans: levels 7 to 10

Level 7

Level 8

Level 9

Level 10

0m 10m 20m

1:1500

housing block is also designed to encourage neighborly interaction by having shared entry spaces to housing that are typically limited to a small number of people, and are uniformly on the building exterior and visible to other residents. The nine buildings each have their own character although several share an entrance.

Placed over the grocery store, De Steven is the tall, twelve-story tower on the southwest corner, with forty apartments—one to four apartments on each floor. From the south, it appears to have two houses perched on top. These are actually two large penthouse apartments with roof garden terraces. Predominantly gray, De Steven has white and red trim with floor-to-ceiling glass windows on the south side. Because of the windy location, apartments facing the east and south have loggias rather than balconies as exterior space, while the few units facing the more protected, courtyard-facing east and

De Zilvervloot floor plans: level 11 to roof

north sides have terraces and balconies. Some of the largest units in the project are in this building, including fourteen three-bedroom social apartments for rent on floors five through eleven.

De Genua is the apartment building on the northeast corner, also on top of the grocery store. Although its gray color is rather somber, at the top of the building on the northwest corner is a projecting box, an extremely large bay window for the sixth-floor living room that lends a little humor and playfulness to the façade. The apartments in De Genua are one- and two-bedroom units that mainly have balconies as outdoor space.

Taking advantage of its entrance being on the commercial street on the north of the site, De Pakhuizen (The Warehouses) contains nine live-work units over commercial space, topped by a four-bedroom

Diagram of the nine buildings

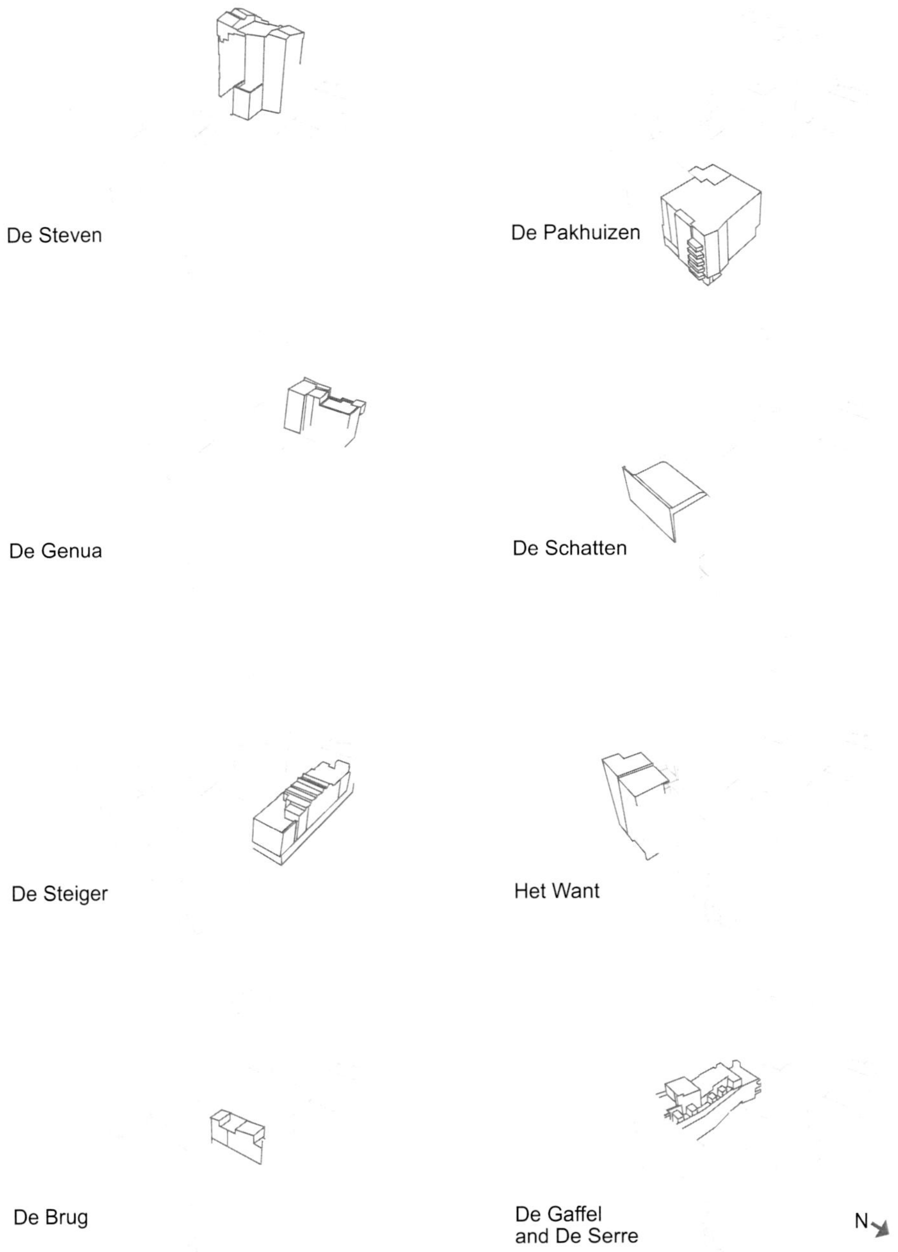

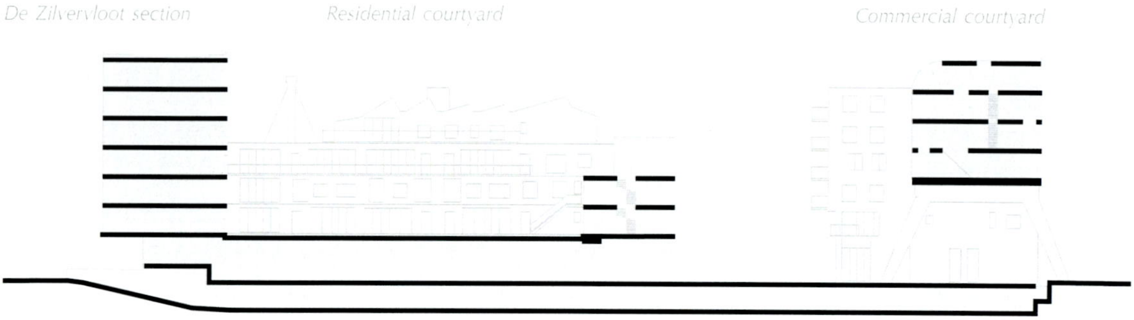

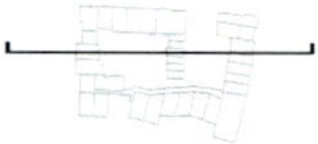

Section

penthouse with a large deck-roof garden. Six of the four-bedroom live-work units are maisonettes, half having access from the residential courtyard-roof garden level, and the other half having access from a gallery overlooking the courtyard two stories above. The last three live-work units are two bedroom flats in a stack at the corner with windows toward the residential courtyard and street, and loggias overlooking the commercial courtyard.

Facing the east side of the residential courtyard on one side and the commercial courtyard on the other are the three De Schatten (Kids) rowhouse buildings. Located above the commercial level, and only 2½ stories tall,

De Genua on the northwest corner

De Steiger and De Brug on the east side of De Zilvervloot

Residential courtyard with De Pakhuizen on the left and De Schatten ahead

these generous three-bedroom plus studio single-family dwellings have a garden on the courtyard level and a terrace on the roof.

Opposite De Schatten facing the commercial courtyard are two connected buildings, De Steiger (The Climb), on the corner, which contains the entrance to both buildings, and De Brug (The Bridge), which forms a gateway between the commercial courtyard and the Admiraalsplein plaza at the east of the complex. While De Steiger is a quiet gray color with red highlights, De Brug is a vibrant red that celebrates its suspension between De Steiger and the adjacent Het Want buildings. Situated over a restaurant, De Steiger is six stories tall with vestibule entrances on every level except the fourth, where the corridor provides access to all nine of the units in De Brug as well as to three apartments in De Steiger. All of the apartments in De Steiger are small one- and two-bedroom units with a mix of balconies and loggias facing east and west. The De Brug units are two and three-bedroom skip-stop maisonettes, reached from unit entrances along the double-loaded corridor that either lead up two more stories or down one level to create the four-tiered bridge. These units have loggias that face either west toward the commercial courtyard or east toward Admiraalsplein.

The east façade of the block combines two solid-color buildings, De Steiger in gray and De Brug in red, with a multi-color building, Het Want (Because). Unlike the other two, Het Want doesn't read as a whole building, but as a collection of units, each with its own red, brown, white, or gray color. But when it turns the corner to the south, Het Want is expressed as a more sober series of vertical elements in red or navy blue. This vertical emphasis on Het Want's east façade contrasts with De Gaffel's east façade, which appears

From west to east (left to right): De Serre, De Gaffel and Het Want form the south façade

predominantly horizontal due to its balconies. De Gaffel reads as an animated set of discrete units, colored in red, gray, or white.

On the southeast corner of the block, Het Want, in addition to containing its own thirteen dwellings, incorporates the elevator access to the adjacent combined building De Gaffel (the Fork), and De Serre (the Conservatory) to become the entry for a total of forty units. The thirteen Het Want apartments are reached from within the elevator vestibule, while the other units are accessed on the building exterior from either the second residential courtyard level, or from exterior galleries on levels 3–5. Eleven of the thirteen Het Want apartments have two bedrooms and one has one bedroom, but the twelfth is a four-bedroom penthouse located on the seventh floor of Het Want that has a large roof terrace.

De Gaffel (combined with De Serre) is a five-level building with four stories of two-bedroom apartments with courtyard-level gardens or south-facing balconies. On the fifth level are three apartments and one maisonette, all with roof gardens on the north and terraces, or in one case a greenhouse to the south or west. Its west end, reached through the galleries, is called the Conservatory because all seven of the two and three-bedroom units have greenhouses as well as balconies and/or roof terraces.

Housing and Unit Typology

Almost all of the other housing projects in this study use clear and defined unit types to organize the building. The approach to unit types taken by AAUI varies across units. Some dwelling types were applied consistently by AAUI. Others follow the ideal of individuality for each unit that requires the concept of unit typology to be applied somewhat loosely, since few individual units will exactly fit the type characterization.

> So nearly 100 units were to be totally different, and the other forty followed three variants . . . For us, the principal ecology was the human one: our answer was the complexity of the human landscape. Honestly, for different families, I feel the architect has no right to impose identical models: it is a crime committed against the expression of a "non-military" society . . . A rule for me is that no apartment shall be identical to any other.[11]

By and large the unit designs are generated by the location of the dwelling, whether at a corner or along an edge, and by the type of access the building location affords, thus following the principles elaborated earlier of light and ventilation on two sides and access to at least one outdoor area. As also mentioned earlier, all units have outdoor space, the exact type determined again by the location of the building and also the orientation of the unit. The variations between units of a given type come not only from size and numbers of rooms, but from orientation (for example, the placement of roof gardens, loggias and/or balconies varies by level and by position relative to sun and wind), geometry (the site isn't rectangular so accommodations in shape create differences between units), and from the type of entry (galleries, courtyard, vestibule, etc.). Instead of showing

De Zilvervloot unit plans

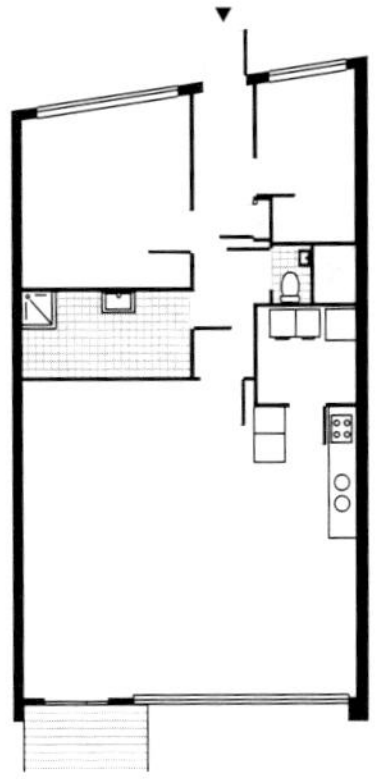

a. South-Facing Flat

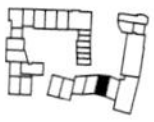

b. Skip-Stop Maisonette

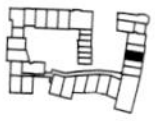

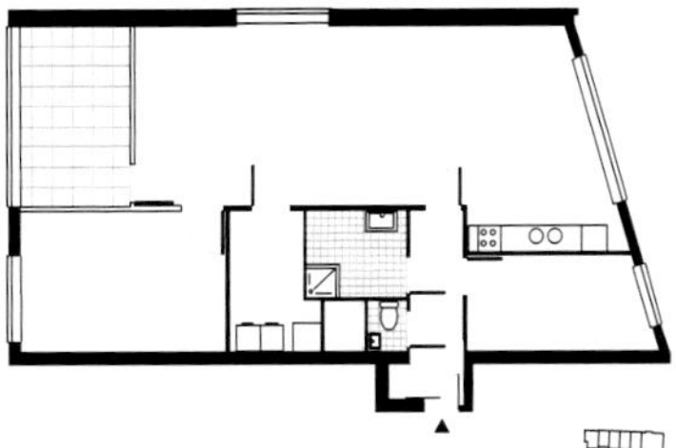

c. Upper-Income

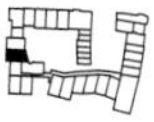

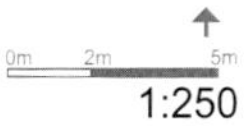

North gallery access for De Gaffel and De Serre, over commercial space

Front door of a De Gaffel courtyard level dwelling

all the unit types, which if one used the real estate brochure to identify would come to twenty-eight, we have chosen a few that give a sense of the range and variety of dwelling designs.[12]

The units in De Zilvervloot are large by Dutch standards, but the organization of the apartments generally conforms to the Dutch concepts of space that, for example, provide an enclosed entry corridor and separate toilet space from bathing space. The designs typically provide one toilet and one bath per unit, even in luxury dwellings.

South-Facing Flat

The south-facing flats are located in De Gaffel and De Serre (see plan a, page 99). Nineteen of the twenty-seven south-facing flats in de Gaffel are social housing, sixteen being similar two-bedroom apartments. These social housing units, oriented north-south, are typically Dutch, with the entry hall connecting all rooms and leading to the open living-dining kitchen space. South-facing balconies are accessed from the living room through glass doors, except on the top floor where the dwellings have roof gardens.

All of the units in De Gaffel and De Serre may use either the Het Want or the De Steven elevator core. The flats on the first and residential courtyard level are reached via a walkway on the north and have their open space, a garden, on the south side. The seventeen flats on the next three levels are accessed from a gallery on the north and most have balconies on the south side. The top level dwellings, one- and three-bedroom variants on the south-facing flats, arrive at their front doors past a roof garden outdoor space on the north. Because the buildings are slightly canted, the gallery isn't straight, and the cants generate front porch areas in front of some units. Furthermore, the galleries also are not precisely one over the other, so that there is play between them, with light coming through vertical openings at certain points.

Kitchen and dining area in the upper-income flat at De Zilvervloot

At De Gaffel all sixteen social housing south-facing flats are stacked four to a level, and at De Serre five such units are stacked two to a level with one at the top. There is no unit for De Serre on the top level. This creates a stepped roof so that two dwellings on the west end have a roof terrace and greenhouse on the roof where the missing units would have been.

Skip-Stop Maisonette

All of the eight apartments in De Brug are two- and three-level skip-stop maisonette units designed for people with moderate incomes (see plan b, page 99). They share a double-loaded corridor and have a view of the commercial courtyard on one side and the Admiraalsplein on the other.

Shown in the accompanying illustration (see plan on page 99) are the four two-bedroom dwellings that enter at the fourth level and go down one level. A visual connection between the two levels is created by an area at the entry that is open to the living room above. One of the bedrooms is on the entry level along with the bath, and the other is on the living room level where kitchen and toilet are located and where a loggia overlooks the Admiraalsplein to the east.

The other four units go up two additional floors. These three-bedroom (or two bedroom and study), three-level units have a two-story living room space on the west side that opens onto a loggia facing the commercial courtyard. The kitchen and dining area are connected, but separated by the open staircase, and face east, with an opening overlooking the entry level below. Two bedrooms are on the top level along with a bathroom that contains a toilet. The third bedroom or study is at the entry level along with a toilet.

Upper-Income Flat

There are a variety of upper-income units at De Zilvervloot, including some penthouses. The unit shown in the illustration (page 99) is a flat in De Steiger, an eleven-story building with forty-four households comprising upper-income, moderate-income, and social housing apartments. These units are stacked, one per level, at the north end of the building. Each of these dwellings is reached via an elevator vestibule it shares with three other units.

These apartments have an unusual design. The largest bedroom does not have its door from the entry corridor, but from the large living-dining-kitchen, even though all of the other spaces (second bedroom, toilet, bath, laundry room and living area) are accessed from the entry corridor. The plans of these units change slightly from one floor to the next in term of fenestration and outdoor space. All the dwellings have windows on the east facing the community courtyard, and on the west facing the main streets, where most units enjoy a distant view over Dordrecht and Rotterdam. Because the top five apartments are not adjacent to but above the shorter De Genua building, they have windows on the north side as well. The photo above shows the dining and kitchen area of this apartment.

The outdoor spaces also vary from one level to the next. All of the dwellings have a loggia in the northwest corner toward the distant view, except the one on the

The top three levels of live-work maisonette (courtyard access, bedroom level and gallery access) with the penthouse above

lowest level that has a garden that opens onto the community space. Additionally, the flat on level 7 has a terrace on the roof of De Genua. This is typical of how the architect takes advantage of variety in building height and unit location to make variations in the residences.

Rowhouse

The only rowhouses in De Zilvervloot, the three De Schatten, sit in the residential courtyard and are reached through the De Pakhuizen elevator core. You enter each of the three large units from the courtyard either by an exterior staircase that reaches the main floor directly, or past a terrace garden into a lower level entry foyer that leads to a small hall with staircase. The large square open living area on the main floor is designed so that the kitchen, and even one additional space, could be spatially segregated with sufficient living space remaining. The lowest level contains three or four bedrooms, toilet, and bath. On the top level, a studio opens onto a roof terrace. The rowhouses have windows toward the residential courtyard on one side and the commercial courtyard on the other.

Live-Work Unit

Six of the ten units in the De Pakhuizen are live-work maisonettes that interlock, so that they have rooms facing both south toward the residential courtyard and north toward the street. Another three units are live-work apartment units stacked at the northeast corner of the building with light on three sides and three or four bedrooms or workrooms. They overlook the street, the commercial courtyard, and the residential courtyard. The nine live-work dwellings share an elevator with a penthouse on the top of De Pakhuizen and with the De Schatten rowhouses. The apartments are reached directly from the elevator vestibule, while the maisonettes are accessed via corridor to either a gallery on the third level that runs along the south façade of the building or via a courtyard entrance past a terrace garden.

All six of the three pairs of live-work maisonettes have bedrooms on the second level. Their plans are identical plans with a very large floor through living-dining-kitchen space, except that some enter on the first level and others on the third level. Those that enter through the residential courtyard level have three bedrooms above, overlooking the courtyard, and those that enter from the gallery have three bedrooms below, overlooking the street. In all cases a loggia that opens from the living room faces south toward the courtyard, and the main floor workspace faces the street. Toilets are located on both floors and the bath is on the bedroom level, whether upper or lower. In this design, with the addition of partitions, the kitchen can become a self-contained space.

Penthouse

There is no "typical" penthouse at De Zilvervloot, but we are discussing the penthouse at De Pakhuizen because it is at the top level, like several other flats and maisonettes, on the roof level. While each of the penthouse units is unique, either apartment or maisonette, designed to address a particular access and roof condition, this dwelling type is important to mention. The penthouses, being on the top floors, have terraces, greenhouses, and roof gardens, as well as a wonderful view over Dordrecht. The penthouse

Roof gardens at De Zilvervloot, with penthouse beyond

at De Pakhuizen (above the live-work units) has an entry foyer, and a long hall encompassing four good-sized bedrooms, a bathroom, and a toilet that leads to a large open living room with kitchen.

Syntactical Structure

When compared to the other housing projects, although the appearance of the building looks more complicated from the outside, the syntactical structure of De Zilvervloot is very straightforward since it only has five points of access from the street and five circulation threads, despite the fact that the project is conceived as having nine buildings. That is because several buildings share entrances: De Pakhuizen shares with De Schatten, De Steiger shares De Brug, and Het Want shares with De Gaffel and De Serre. While the five threads are quite discrete, they each have a distinct character, indicating the variety of ways that circulation is employed in the project.

The syntax diagram also shows that the building is relatively integrated or permeable on the interior. In addition to the sharing of entries, all of the dwellings are connected to the community residential courtyard, albeit a little indirectly for De Steiger. In terms of fan circulation spaces that connect directly to units, whether

De Zilvervloot space syntax gamma diagram

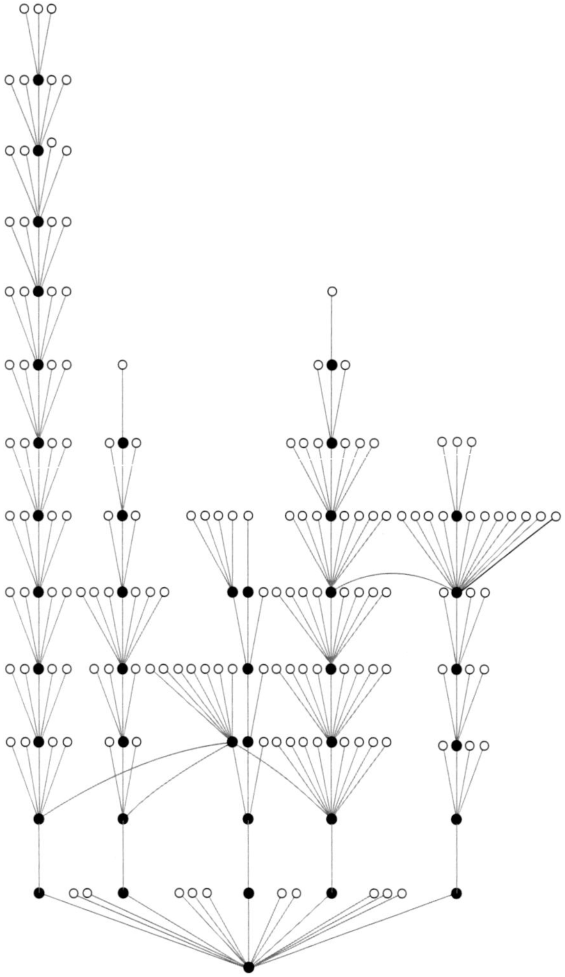

However, the building is not permeable to the exterior. The syntactical structure supports the security of the building through the distinct separation of the commercial and dwelling places, from the few, clear access points and also by means of limiting the number of units at each level that access the elevator core to increase neighbor recognition. The main complexity in terms of linkage comes from having all buildings linked to the courtyard level, and from having connections between four of the buildings on levels 1–4.

Social Principles

The nine separate but related buildings that form the urban block and the two courtyards each have a different mix of dwelling types and a different form (gallery accessed, vestibule accessed, double-loaded corridor). The ground-level commercial plaza that also serves as entry to two supermarkets is accessible to the whole neighborhood. The second residential courtyard, located on top of the supermarkets, is accessible only to the residents, and serves as shared open space for the complex.

The Wielwijk planning process emphasized the importance of having participation with a diverse community that would include a variety of income levels and family types. The architects were called upon to create a mix that would work. The many different kinds and sizes of units within the complex discussed earlier were designed to attract and maintain a variety of lifestyles and income levels.

vestibule, corridor or gallery, there is only one place in the building where more than seven units connect to a single access point. There are an unusual eleven units where the De Brug maisonettes connect to De Steiger, creating an additional internal ring. Opportunity for neighborly interaction is supported by the overall interconnectedness of the units with the residential courtyard level.

The combination of units within a given building is not consistent across the buildings. Because there is no legible pattern, with some exceptions residents are unlikely to know exactly which units are social rental unit, which are moderate income for purchase, and which ones are luxury units. For the most part this housing complex does not segregate by income even within buildings. Although the standard division for new housing in the Netherlands at

Commercial street on the north with De Pakhuizen entrance and façade

Entry and mail area for Het Want, De Gaffel, and De Serre

the time was 33% social housing (rental) and 67% moderate income and high income,[13] here the division is 50% social and 50% moderate and luxury. Previous to this project, various housing corporation projects had successfully combined high and moderate and moderate and social housing. But here there was concern because the architects wanted to mix all three together in a single building, sometimes having a single luxury unit in a building with predominantly social housing. Woondrecht, the housing corporation, was concerned that there might be no buyers for the luxury units.[14]

However, contrary to the worries, the luxury apartments were the first to be sold,[15] perhaps because many of them are penthouses at the top of the building with roof gardens, most have excellent views, and most share a vestibule with only a few units. Sharing an elevator core and mail room with a mix of people did not seem to pose a problem for high-income residents in the first decade of the 21st century in Dordrecht.

None of the buildings has exclusively social housing. If we consider those buildings that share an entry and have all three income levels, for example, (1) Het Want, De Gaffel and De Serre (40 units), and (2) De Brug and De Steiger (23 units), as well as (3) De Steven (40 units), we can see that 103 of the 130 units are in a mixed-income building. Of the remaining buildings, (4) De Genua combines five luxury flats with nine social flats, and (5) De Pakhuizen and De Schatten, which share an entry, combine nine moderate income units with four luxury units. So the mix follows no clear pattern. It will be interesting to see if any problems related to this issue arise in the future.

Perhaps one reason the mix works is that the design of the entries and access corridors supports the natural and informal surveillance of neighbors and visitors. With the exception of the De Steven, the number of households entering from the street at any given elevator access point and mail reception area is limited to less than twenty-three, less than what is found a maximum size for maintaining a community without requiring specialized security.[16] And for De Steven, the possible negative effect of having forty-three people enter the building is reduced because there are no more than four households on each level, allowing people to be very familiar with the few neighbors on their vestibule or corridor. Furthermore, all vestibules areas in De Steven (and most in the complex) are naturally lit, in part due to the Dutch building codes, which as noted earlier, unlike those in the United States require access to the outdoors prior to entering an emergency stair to alleviate potential smoke.

View of De Zilvervloot from the southwest

For almost half of the De Zilvervloot units, corridors are eliminated altogether in favor of vestibule access (De Steven, De Genua, Het Want, and De Steiger). But there are some relatively long distances between elevator and unit (normal, however, within the Dutch context). The longest and only unlit corridor is in De Brug, where the twelve moderate-income maisonette units share a hall that is approximately 120 feet (or 40 meters) long and connects to the lit vestibule access point of De Steiger. Additional long access routes are the galleries that connect the Het Want elevator core to the eight or so units on each of the first, second, third, and fourth levels. The final long routes are those that connect the elevator access point of De Pakhuizen to the live-work maisonettes and to the De Schatten units either at the residential courtyard level or from the gallery at level 3.

Lessons and Conclusions

While the long-term success of the Wielwijk social experiment remains to be seen, the short-term success of De Zilvervloot seems clear. One lesson is that engaging in participatory design with a neighborhood can lead to neighborhood solidarity that supports positive urban transformation. Today the neighborhood that had contained only social housing has a mix of incomes, commercial venues that are viable, and fully occupied dwelling units with gardens, terraces, and loggias personalized by the inhabitants. The introduction of commercial activity was essential to the success of the project because it engenders ongoing activity that enlivens the neighborhood.

The project with its variety of design arrangements also begs exploring a variety of questions about such things as how the mix of people within a given building, and the number of people sharing an elevator core, vestibule, or gallery affect both the satisfaction of the residents and the ability to sustain the building's vitality. But the building was completely sold and made a good profit for the developer.[17]

An important contributing factor to the success of De Zilvervloot is the careful and continuing oversight of the housing corporation. Because the rental income makes the project economically viable, the corporation has an interest in ongoing success and maintenance. For example, rather than allow the local supermarket tenant Albert Heijn to flounder by simply following its generic approach to grocery sales, the developer worked with the supermarket tenant, Albert Heijn, to customize the groceries to match the needs of the local ethnic populations, thus ensuring its commercial success.[18] To have thriving neighborhoods and successful housing, designers, planners, and developers must engage with the community and be motivated to make long-term investment rather than take short-term profits.

Notes

1. Rob Hagens (Director Stichting Woonbron, developer), Interview, Dordrecht, March 2, 2010, 1.
2. Ibid., 1–2.
3. Lucien Kroll (Atelier d'Architecture, d'Urbanism et d'Informatique, architect), Site Visit, Dordrecht, March 2, 2010.
4. For information on Open Building, see Stephan Kendall and Jonathan Teicher, *Residential open building* (London: E&FN Spon, 2000).
5. As noted in Wijken Dordrecht, "Nieuwsbrief Wielwijk," (January 2016), 3, www.wijkendordrecht.nl/wielwijk/projecten-in-de-wijk/nieuwsbrief-wielwijk-met-alle-plannen-voor-de-wijk; and Gemeente Dordrecht, *HOOFDSTUK 3 Planschrijving nieuwe situatie. ND*, www.ruimtelijkeplannen.nl/documents/NL.IMRO.0505.BP155Herzwestervoe-3301/t_NL.IMRO.0505.BP155Herzwestervoe-3301_0002Planschrijvingni.html.
6. Rob Hagens, "Pourquoi Wielwijk?," unpublished PowerPoint presentation, Woondrecht, 2008.
7. Lucien Kroll, Site Visit, 2010.
8. Lucien Kroll (Atelier d'Architecture, d'Urbanism et d'Informatique, architect), Interview, Dordrecht, March 2, 2010, 5.
9. Lucien Kroll, "Atelier d'Architecture, d'Urbanism et de l'Informatique: Zilvervloot," *A+U: Architecture and Urbanism* 6, no. 429 (2006), 85.
10. Lucien Kroll, Site Visit, 2010.
11. Kroll, "AAUI: Zilvervloot," 84–85.
12. Woondrecht Vastgoed, Ooms Drechtsteden Makelaars, and Waltman & Company, *Zilvervloot: A Living Dream* (Real Estate Sales Document), Dordrecht, 2005.
13. Eric Amory (De Key, developer Silodam), Interview, Amsterdam, May 25, 2012, 3, and Gerard Comello (Lingotto, developer Whale), Interview, Amsterdam, March 5, 2015, 1.
14. Lucien Kroll, Site Visit, 2010.
15. Ibid.
16. Julia W. Robinson, *Institution and home: Architecture as cultural medium* (Delft: Techne Press, 2006), 167.
17. Twelve percent profit including both the residential commercial properties, according to Rob Hagens, Interview, 2010, 2.
18. Rob Hagens (Director Stichting Woonbron, developer), Conversation, Dordrecht, February 22, 2015.

5 Carnisselande

Carnisselande south façade with clock tower

Location of Carnisselande in the Netherlands

Location of Carnisselande in Barendrecht

Location of Carnisselande within its neighboring housing and the Middeldijkerplein

Carnisselande on the Middeldijkerplein, the central square of Barendrecht, is a landmark for the new part of the town. At the time of the development of this suburban area outside of Rotterdam, the mayor of Barendrecht envisioned the project as including a church tower, the traditional signal for the presence of a town on the flat Dutch horizon. However, with the decline of religious activity in the Netherlands, no church was interested in new construction. The mayor, still committed to the idea of a spire, decided that the presence of the central square of Barendrecht, site of the shopping mall, police and fire stations, and town administration in the countryside, could instead be marked by a clock tower comprising housing.[1]

Originally a typical small Dutch town, Barendrecht was designated by the city of Rotterdam as an area for suburban development. At that time, the town residents, seeking to retain the identity of the original town, decided to create the new development as an independent area. Thus the center of the new suburban area was created around the plaza.

The Carnisselande project houses the full range of incomes at a density of 190 dwellings per hectare (80 units per acre). Most of the residents have come from Rotterdam, finding in Barendrecht a good place to live, with a twenty-minute tram ride to the center of Rotterdam, and easy access to the surrounding countryside, a polder area with farmland, and many bicycle trails.[2]

Amina Fatehmahomed moved to Carnisselande from Rotterdam, seeing it as a good place for a single mother to raise her children, and so that her daughter could have her own room separate from her two brothers and her mother. She has found living in Barendrecht to be very affordable and has been able to pay off her debts and to save money since she arrived. At the time of her 2011 interview by the

The tower marks the center of Barendrecht from a distance

The main façade of Carnisselande

periodical *Via Vestia* she had lived there for five years, having moved from a three-room, two-bedroom apartment in Rotterdam to a small house in the project. Her children were traveling to school by bike, and she often had coffee with a neighbor friend who is also a single mother. Happy with her new housing with its garden and extra bedroom, Ms. Fatehmahomed found that Barendrecht is quieter than Rotterdam, maybe even a bit boring, although she could take the children to the cinema and out for breakfast and shopping. She often returned to Rotterdam for the Saturday market where she felt totally at home. Her move to Barendrecht was to benefit her children: "I moved here for them, I want them to do well."[3]

Site, Context, and Building Functions

The Carnisselande project, designed by Steenhuis Bukman Architecten and developed by Estrada Projecten in 2006, includes a full range of housing,

The Middeldijkerplein
© Jannes Linders

including a group home. Of the 187 units, about half were intended to be for rent and half for sale. In the end, however, more units were sold than rented as there was great interest in living in the area, and the number of rented units is between twenty and thirty (about 13%).[4] As alluded to earlier, the project frames one side of Middeldijkerplein, the main plaza of the new area Barendrecht, with eight-story buildings on either side. These three sides of the square comprise buildings with civic functions and business fronting the plaza and housing above. Opposite the project on the fourth side is the shopping center with a variety of stores and services. The tram to Rotterdam crosses the plaza diagonally and stops in the middle, providing constant activity on the central square.[5]

The Carnisselande complex includes housing, parking (both underground and at ground level), a clinic, a library, and several businesses. The main block of housing faces the square alongside the housing tower, with the remaining housing branching off in the back, in four wings that create two small garden courtyards, each enclosed by a final block of rowhousing. Because the site slopes, the plaza in front is one level above the canal level behind. Thus the parking under the main building can be reached from the ground level at the rear, with the remaining parking located on grade at the lower back part of the project. Storage areas for the flats, large enough for bicycles, are located at the base of the building, or adjacent to the underground car park. Rowhouses have sheds in the garden to store bicycles and other items.

The housing wings, the two courtyards, the onsite parking, and the pathway to the adjacent housing area

The passage to the square: library on the left and clinic on the right

A path going underneath the Carnisselande complex links the Middeldijkerplein to the housing area behind. The gateway below the main housing block also serves as an entrance to the clinic and the library. Once beneath the building, the pathway leads to a stair to the lower exterior parking area, continues through a green area along the parking, crosses the rear parking area, and bridges the canal to the extensive housing areas on the other side, connecting the central square and tram to the housing developments beyond.

Carnisselande floor plans: ground level and level 1

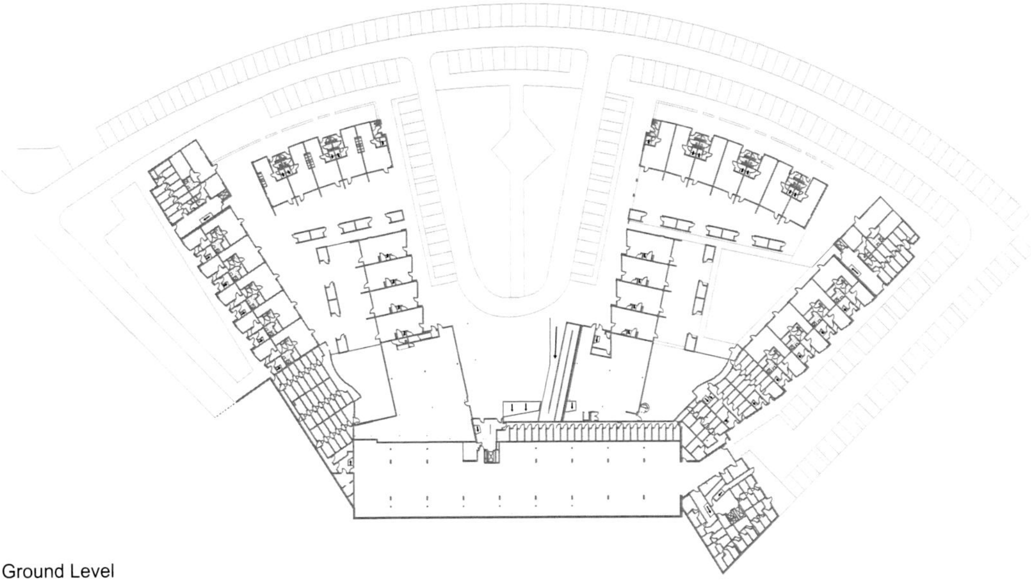

Ground Level

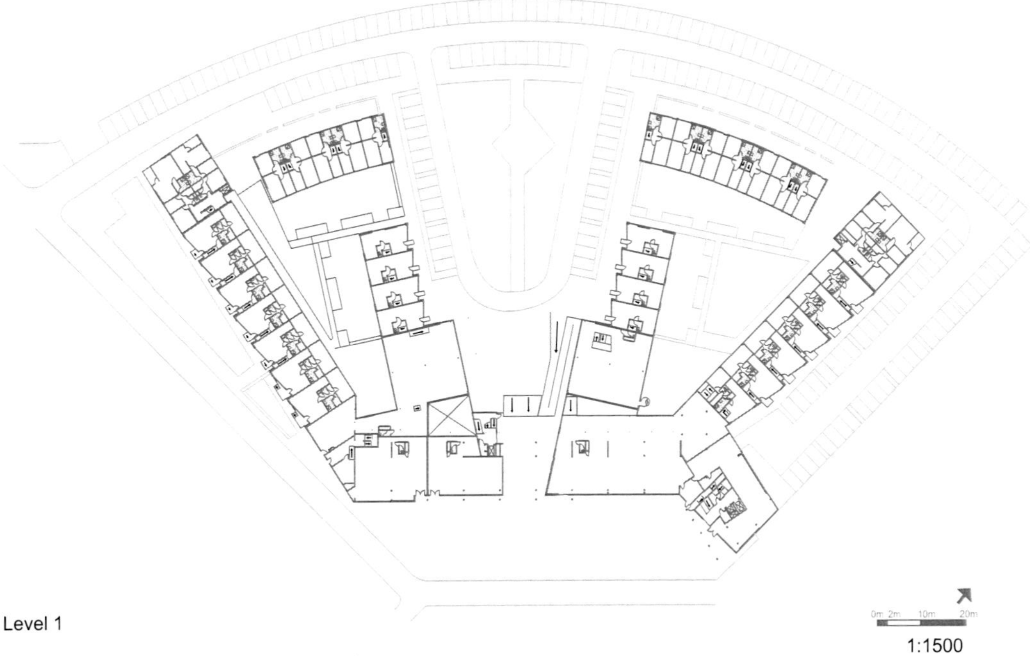

Level 1

0m 2m 10m 20m

1:1500

Carnisselande floor plans: levels 2 and 3

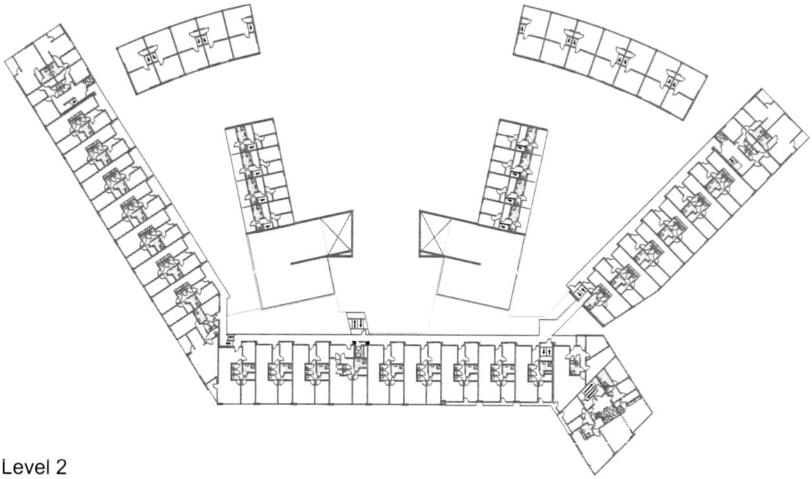

Level 2

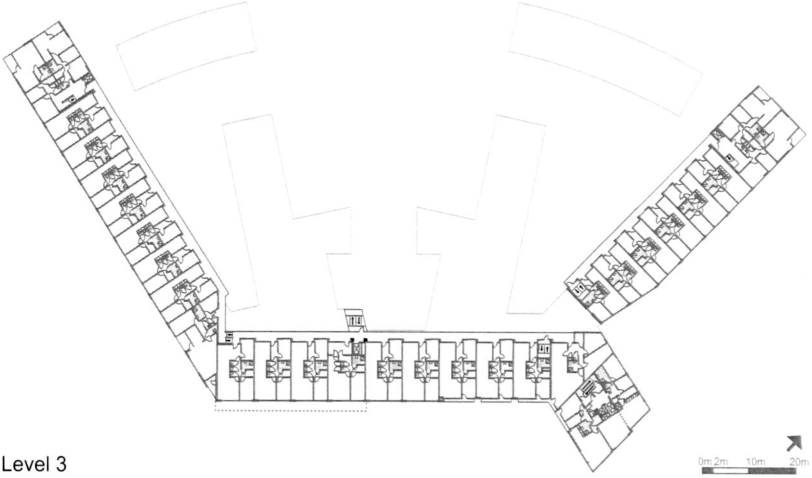

Level 3

1:1500

Carnisselande floor plans: levels 4 and 5

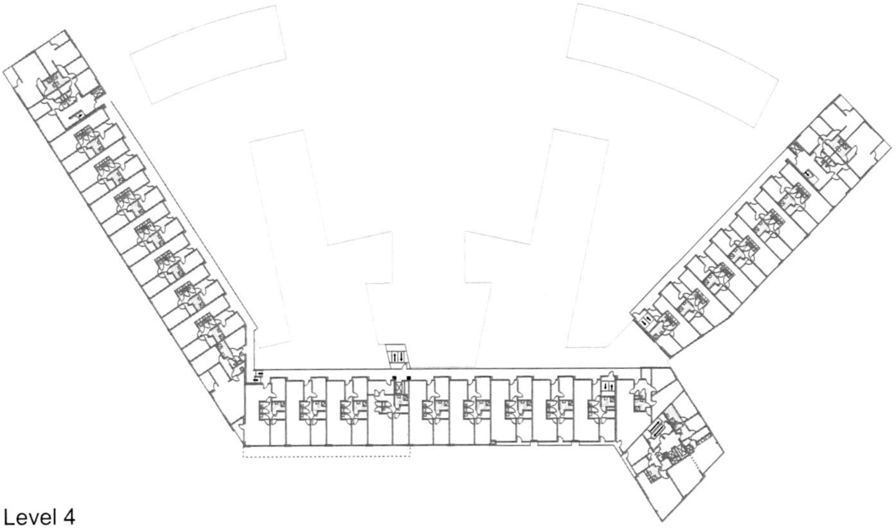

Level 4

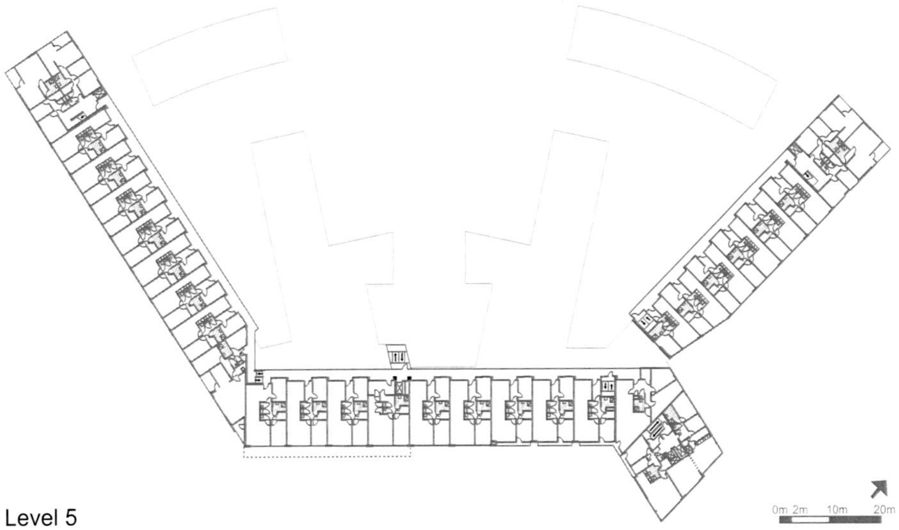

Level 5

1:1500

Carnisselande floor plans: levels 6 and 7 and levels 14–18

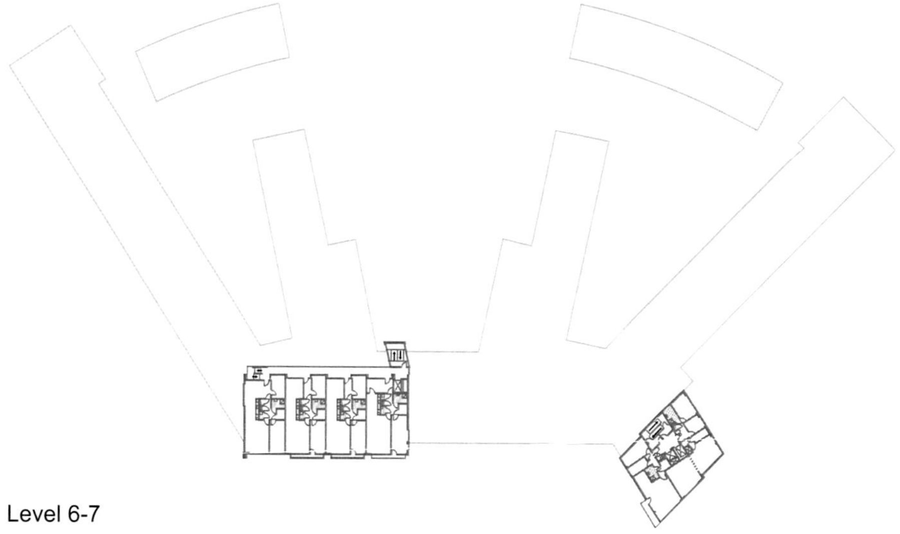

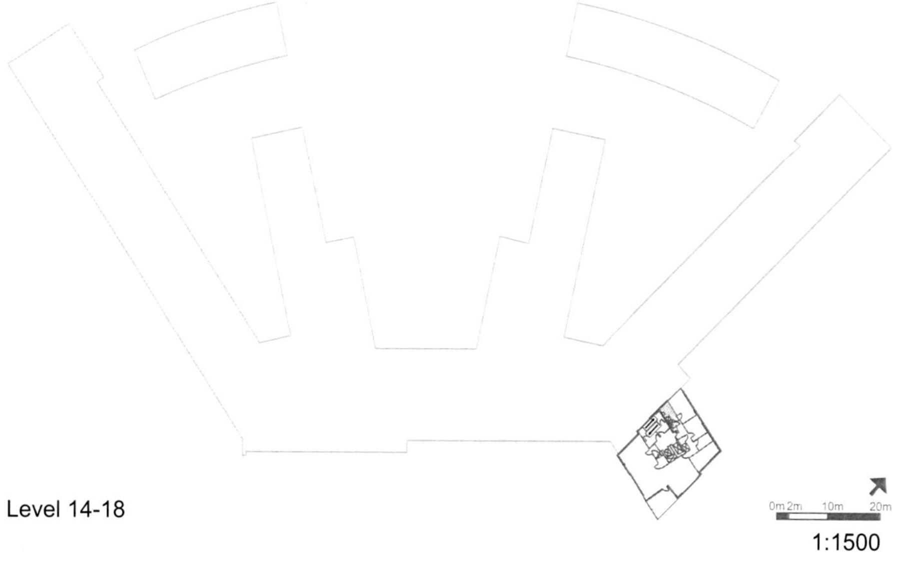

The three-story rowhouses on the north relate in scale to the adjacent residential area

Organization, Appearance, Form, and Massing

Carnisselande's massing has a strong relation to the larger urban pattern, with the taller front section tapering off to three stories at the canal. The eighteen-story tower and the five- to seven-story main block frame the square; the open passage at the base of the building links the town square to the blocks beyond; and the three-story blocks at the rear match the residential scale of the adjacent area.

Because the Middeldijkerplein is oriented at 45 degrees north and south, the project's main block and façade faces southwest, and the four wings can be called West, Middle West, Middle South and South (see diagram on page 120). The two garden courtyards are here designated West and East and the courtyard-enclosing rowhouse sections North and Northeast. Each section of the building has its own identity and typically includes one or two housing types for a particular income range.

The project is skinned in a cheerful orange brick with white concrete and white panel accents. The main block is expressed as a variety of forms including a cantilevered area with deeper apartments surfaced with white panels (see photo page 127). The windows of the gallery flats on this façade vary from expression as punched individual windows along much of the main face, to ribbon windows at the top level that accentuate the horizontality of the block in contrast with the vertical tower. The overhang and shadow line formed by the setback of the ground level commercial area give human scale to the large block. The gateway invites movement through the block to the clinic, library, and lower parking area. The housing in the block consists of flats designed for sale to middle-income residents.

The tower at the south corner, paid for by the housing corporation, was carefully studied by the architect to create a sculptural form, with the 16-meter dimensions making it the most slender housing tower in the Netherlands.[6,7] Its dimensions, its diamond shape, the projecting roof, and the use of glass in the corners on

The Carnisselande building diagram shows the relationship between courtyards, tower, main block, and wings

The Carnisselande tower: lower income flats at the base, middle income flats in the middle, and luxury flats on the top levels

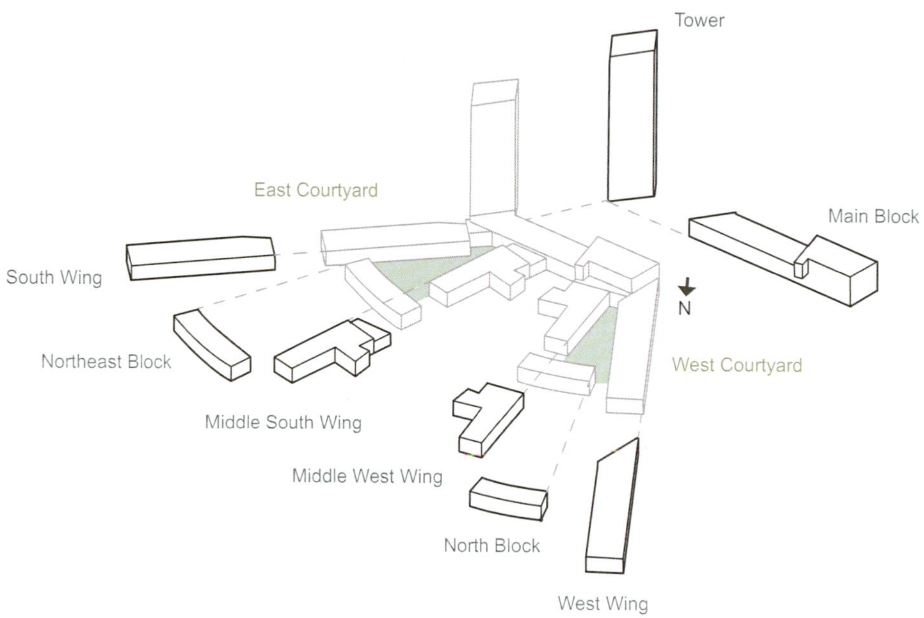

certain floors give it a light and soaring appearance. The clock on two sides is visible for miles around. Within the tower, levels 2–5 consist of three lower income flats, and levels 6–13 comprise pairs of middle-income flats, with five luxury units at the top, one per level.[8] All of the units in the tower were sold and the owners have their own homeowners' organization, responsible for maintaining the building, the elevator, and controversially, the clock.[9] At the base of the tower is the entrance to the interior parking that was designed to serve the main housing block, not the tower, but since there are fewer cars than anticipated, spaces are available for rental to other residents.[10]

Each of the four wings at the back of the building is slightly different. The West Wing was designed to be entirely social housing, comprising two levels of rowhousing at the base with three levels of gallery-accessed flats above. The West Wing street façade is relatively plain, with the rowhouse unit entrances articulated by a small setback that creates a shadow at the lowest level. Three of the flats on the plaza end and four on the canal end have loggias, balconies and/or cutouts that create a counterpoint negative space, while the rest of the façade is made of sets of three rectangular windows, framed together, one per dwelling. The back façade of galleries with glass railings is visible from the parking area because the wing is taller than the adjacent townhouses. The social housing rowhouse units at the base of the wing that enter from the street side do not have their own gardens, but have a small patio area that opens onto the enclosed communal courtyard that serves the entire complex.

The West Wing façade: rowhouses and flats

The West Courtyard: community garden, outdoor terrace for the West Wing flats, and gardens for the Middle West Wing and Northwest Wing rowhouses

Opposite the West Wing is the Middle West Wing formed by a line of rowhouses next to the clinic that were designed as live-work dwellings. These now serve the group home and as social housing for several large families. These units have gardens at the back enclosed by a brick shed and hedges to form the edge of the communal area, as well as balconies at the top level. Again primarily brick, these four dwellings have a two-level angled setback of glass windows and door to delineate each entrance, and at the top, a horizontal set of two windows framed together with a white panel in-between.

Enclosing the West Courtyard is the North Block—five large rowhouse units planned for sale to middle income residents, designed with a low brick wall to define a garden at the front, and a brick shed and fence to define the garden at the rear. Set back from the street behind the garden wall, the façades of the dwellings are relatively plain, with recessed entries paneled in white, a horizontal band of windows at the top unifying the row,

The Middle West Wing (similar to the Middle South Wing) showing units originally designed to be live-work

The Northeast Block of rowhouses (similar to the north block) with front gardens

Carnisselande from the bicycle path: the South Wing

and at the middle level, square, full-height windows that identify individual houses. This row of five houses is visually connected to a similar set of seven rowhouse units that form the Northeast Block of the East Courtyard. Because these twelve units were sold with the land beneath, the owners are not part of either of the two homeowners' organizations, but have created their own informal group to address joint concerns.[11]

The Middle South Wing, adjacent to the library, is similar in design to the Middle West Wing as it was designed for and serves as live-work units. The two sets of dwellings have a similar façade design and garden relationship, although they are oriented in opposite directions.

The South Wing that forms the third side of the East Courtyard comprises four rowhouses at the base along the west end with two layers of eleven flats above, and a set of seven stacked flats at the east end. The design of this wing of the complex, visible from the bicycle path, focuses on the corner stacked flats, defined by a cutout area of white panels and balconies that contrasts with the brick façade of the rest of the wing.

The remaining section of the South Wing is more sedate, with simple horizontal bands of windows framed together at each apartment, and two-story framed glass and white-paneled openings with small balconies defining the rowhouse dwellings. The opposite side of the wing, like that of the West Wing, has galleries with glass railings on the flat levels with the rowhouses being expressed by glass doors at the ground level with framed horizontal windows and white panels above for each unit.

Housing and Unit Typology

Carnisselande consists of five housing types (flats, rowhouses, maisonettes, live-work units and group home) with nine unit types, with some variations for such situations as corner or end units: social housing flats, social housing rowhouses, middle-income rowhouses, live-work units, group home, lower-income flats, middle-income flats, middle-income maisonettes, and luxury flats. As indicated by the earlier discussion, the pattern of residences within the complex is highly organized, with different wings associated with one or at most two income levels and particular unit types.

The Plaza Block comprises five levels of middle-income gallery flats over commercial and civic space and parking at the bottom. Some of the flats are longer in dimension (those in the white overhanging block), but all have the plan arrangement of *typical flats* (see the next section). The West Wing was designed to be completely social housing (rental), with two-story rowhouses at the base, one-bedroom corner flats at the north ends of each level, and the rest of the units being typical flats with gallery access, although as mentioned before, much of this housing was sold.

The Middle West Wing and the Middle South Wing both were originally designed to be live-work units with a full story of workspace on the bottom level, living space on the middle level, two bedrooms and balcony on the top, and a garden defined by a brick shed and hedges to the back. But in the end, those in the Middle West Wing were not used for work as planned. Two of these houses were transformed into a group home for twelve residents (now being eliminated in favor of smaller residences dispersed in the community), and the remainder were deployed for extra-large families. It is not clear whether the Middle South Wing units function as live-work units.

Rowhouses for sale to middle-income families were the choice for both the North Block and the Northeast Block. Rowhouses are also located at the base of both the South Wing and West Wings, with corner unit flats stacked at the east end and typical flats above the rowhouses. The tower's units are unusual with pairs of flats forming the main midsection of the building, and luxury flats above.

Typical Flat

Typical flats with gallery access are found as social housing in the West and South Wings of the building, and as middle-income housing in the central block. These units are called typical because they make up almost half of the total units in the complex, and have the basic, typically Dutch layout with a controlling entry hall that serves both social housing units and middle-income units—those in the overhanging white block having an additional meter or so in length. Furthermore, this plan is very similar to those of other projects, with the layout as described by architect Roelf Steenhuis,

Carnisselande unit plans

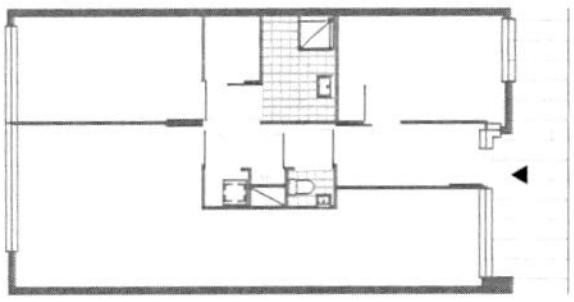

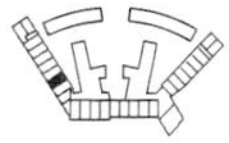

a. Typical Flat

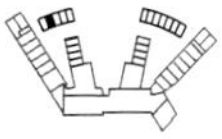

b. Rowhouse Unit

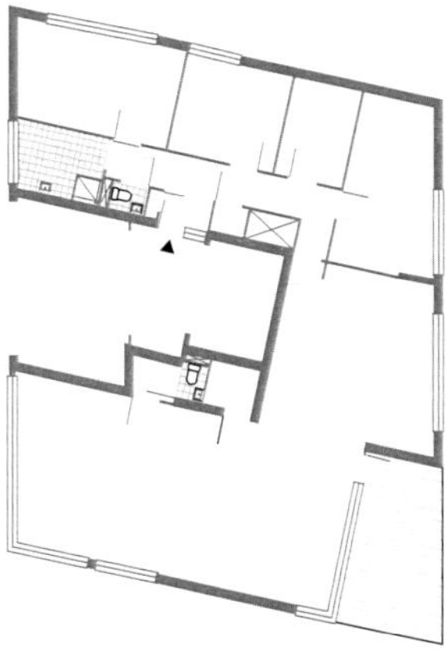

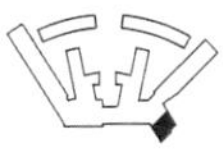

c. Tower Luxury Flat

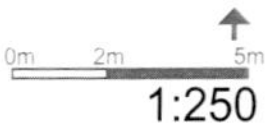

encompassing a single structural bay that permits moving of internal walls. As he points out, this opens up the possibility for the unit to have a living room and kitchen with three additional rooms that can be three bedrooms, or some combination of bedroom and spaces such as dining room or study. Additionally the doors can be placed in several different positions to support handicapped access or other needs.[12]

By and large the typical units at Carnisselande lack a balcony, although several of the middle-income units include a partial balcony. The lack of full balcony space is due to the complex having been built in a short period when the requirement for balconies was suspended and the developer chose to omit the balconies or to build partial balconies instead. The partial balconies are only wide enough to stand on or to have a single chair.[13] Instead of balconies, the residents use the galleries for outdoor space, as there is just enough room for sitting. In fact, many residents have put chairs there, which makes the galleries appear inviting, but they are not often used since most of the galleries are on the north, shady side of the building.[14]

Koos and Gonny van der Elst's apartment is one of the extra-long typical flats. Former residents of Rotterdam, they chose to move to Barendrecht for its easy access to Rotterdam and the extensive bicycle trails. They were seeking housing all on one level, which is what pleases them the most about their flat. Having bought their unit new, they chose to have a dining area instead of a third bedroom. They are very pleased with their dwelling, finding it spacious and accommodating and are pleased to be on the very top floor with no one on top of them. They especially appreciate the view, and enjoy watching the activity on the Middeldijkerplein as well as ships that pass by on a main transport canal also visible from their window.[15] Generally contented with their flat, they had two reservations. First, the window design impedes the use

Looking from the dining room through the kitchen to the living room in the van der Elst flat

of a proper curtain, and second, the dwelling lacks a real balcony and they would like to be able to eat outside: "That is what I miss—a balcony. That's the only thing we miss." But overall they are very satisfied: "We have nice space for two of us. It is perfect."[16]

Rowhouse

Designed for families, these brick three-story dwellings form two curved blocks at the back of the complex (see plan b on page 123). These generous units are oriented east-west, with the main living space on the ground floor. The entrance hall is a traditional Dutch style that controls access to the rooms and stair. It opens into the kitchen with a dining area toward the street and the living room opening onto the garden. Three bedrooms and bath are located on the middle level with another bedroom, bath, and extra living space on the top. The extra room faces the street and canal. In some of the units one of the street-facing bedrooms has a small Juliet balcony. As noted earlier, along the sidewalk in front of the unit is a low wall that defines a good-sized garden that residents have taken advantage of with plantings. Also noted is the presence of a back garden that opens onto one of the two communal gardens for the complex and is framed by a shed and wooden fence or hedge.

The northeast block family rowhouse garden with the gardens and balconies for the middle south units beyond

The living and dining areas of the Onder de Lindens' luxury flat open onto a terrace (visible beyond)

Tower Luxury Flat

There are five luxury flats in the Carnisselande tower, each of which encompasses an entire floor (see plan c on page 122). Although the plan that is shown is for the lower four units, the photograph is taken in the top unit, which has a similar floor plan, except that the entrance was changed to come through an entry into the living room rather than the bedroom corridor.

The layout of these units varies because the first owners chose to arrange things uniquely. But the original plans called for an entrance into the main corridor that first passes the bedrooms and ends in the living area. The large living-dining area faces south with windows on the east and west. The kitchen faces east and the three additional rooms face north, although many rooms have windows on two sides. In addition to the interior spaces, each unit has a large terrace on the southeast corner.

Gerard and Wil Onder de Linden have inhabited their unit for several years, having moved from Rotterdam for reasons similar to those of the van der Elsts:

> If people get older and they cannot drive a car any more they can still live here, so it is very convenient for older people. But also for the youngsters because by bike you can go anywhere from here . . . alongside the river is nature, and also biking roads through this nature area. So you can bike here for hours.

The Onder de Lindens are content with almost everything, except what they see as unnecessarily large windows because they are expensive to replace and thus expensive to insure. However, they appreciate the high insulation qualities in the winter.[17]

Another issue the Onder de Lindens raised does not have anything to do with the unit design, but has to do with the homeowners' organizations. They are concerned that there are two organizations within the complex, and that the smaller one for the tower has only 30 owners (versus 121 owners in the larger organization) but has to maintain the tower clock that serves the city and the whole project.[18]

Syntactical Structure

The organization of Carnisselande is difficult to read from the plan because the building heights are only implied, and the separations between the connected wings aren't immediately obvious. Furthermore the project is quite

Carnisselande space syntax gamma diagram

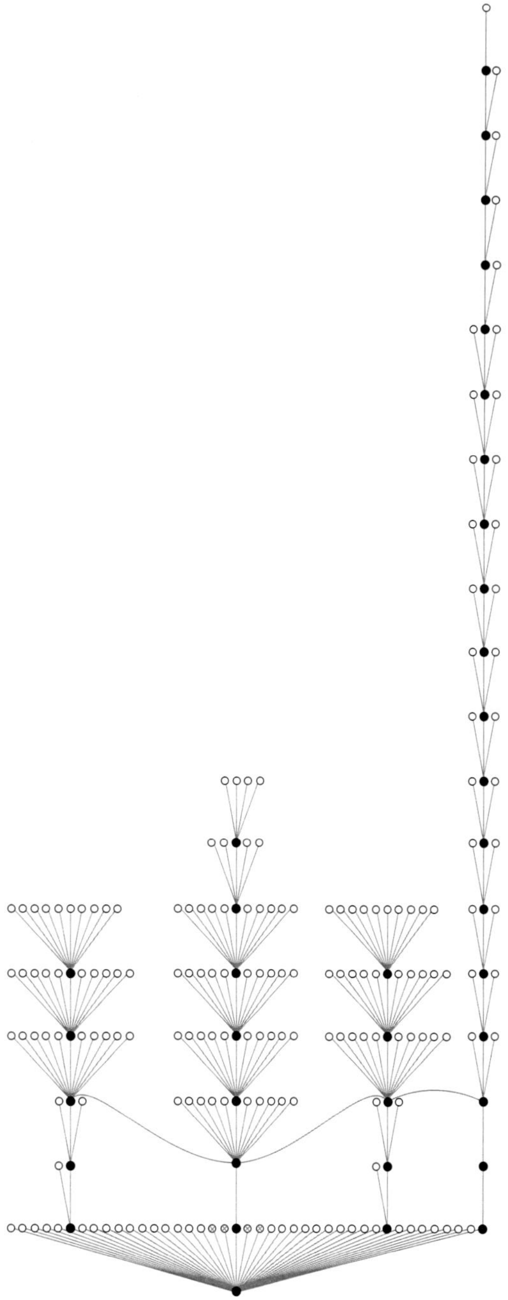

large (the second largest number of units among the eight cases studied). However, Carnisselande's gamma diagram shows a well-organized and straightforward structure. Connected at the first floor for fire exiting purposes are four major threads representing the four wings that have elevator access: Main, West, South, and Tower. In the sense of having a clear structure, the project is comparable to De Zilvervloot. However, unlike De Zilvervloot, this project has many housing units that are reached directly from the street, which is evident from the many connections at the base of the diagram. The use of rowhouses on the lowest levels of the wings, the incorporation of wings entirely formed of rowhouses, and the inclusion of civic and business enterprises at the ground level generates activity on the street.

The connections at the first floor for fire purposes and for access to the communal outdoor area are also similar to De Zilvervloot, providing a certain degree of connection within the project that could be seen as a possible security problem. However, because the primary circulation system is galleries, and the activity on the galleries is highly visible from many points within and without the building, this provides a higher degree of security than would the use of double-loaded corridors that are invisible from the outside.

Social Principles

While the original intention was that Carnisselande should have a high proportion of social housing, the political context in the Netherlands changed greatly between its inception and its completion. As a result of recent policies that require housing corporations to sell a certain proportion of their social housing portfolio, a majority of the units designed to be rented as social housing were sold on the open market to middle- or lower-income residents, perhaps some who had been previously renting social housing and paying market rate. Thus instead of being a project that was evenly split

Carnisselande main façade
© Jannes Linders

between upper, middle and social rental housing, the project is now approximately 20% upper income for purchase, 60% middle and lower income for purchase, and 20% social housing.

Integrating social housing with middle- and upper-income housing is sometimes seen as controversial in the Netherlands. At Carnisselande, the social housing was designed to be located in its own housing wing with its own entrance. Additionally, several large family units and the group home were located side by side next to the clinic with their gardens part of the North courtyard. In the tower, several floors of luxury units were combined with floors of lower and middle-income units and share an entrance. The middle-income rowhouses and live-work blocks were designed to be sold and to be independent of the rest of the complex. The design thus gave each income level its discrete section of the complex (the one exception being the tower that combined upper, middle, and lower-income residents), with mixing taking place in the public spaces and shared courtyards. The design follows the principle of limiting the number of people on a circulation space, here the maximum being ten residents on a gallery, and also putting an important presence on the street with many rowhouses and businesses having entrances there.

In discussing the effect of the social mix on the satisfaction with the four residents I interviewed (two couples), I found that the Onder de Lindens who owned the tower apartment and were part of the tower residents' association couple were not even aware that there were social housing units in the complex. However, Koos van der Elst (chair of the larger residents' association that includes the developer, Vestia, to represent the social housing residents), said that overall the complex runs smoothly, but that "We have difficulties with about eleven dwellings." From his follow-up discussion it appeared that most of these problem dwellings were social housing for large families and the group home, and that the problems tended to occur in the outdoor areas, the communal gardens and on the street (such as loud parties and barbeques sending smoke up to other people's units). Keeping the large families with lots of children and group home together in one area seemed to augment the difficulties. Thus some of the problem may be alleviated by plans to move the twelve group home residents to several smaller residences of three or four people and disperse the dwellings within the broader community, perhaps having one or two at Carnisselande. Perhaps at least one of the live-work units now used for the large families should be sold for its original purpose. While there will always be disagreements between residents, the plan to limit the number of potentially problematic residents located in any one area of a project seems a wise approach.

Achieving a mix of lifestyles and incomes in a project is a challenge. The designers attempted to achieve this by segregating most of the units by type so that incomes wouldn't mix in internal circulation paths. Of course in the end, the sale of most of the social housing resulted in a combination of incomes across the project, although, with the exception of the tower, not the highest and lowest incomes in any one building.

Lessons and Conclusions

This project offers several interesting lessons, one of which we just discussed, the possibility of dispersing residences with larger numbers of people or residents with challenges across a complex, rather than concentrating them in one place.

Another lesson to be taken from this project relates to the homeowners' organizations. While the motivations for having two homeowners' organizations seemed to create a sound division between areas that have higher and lower maintenance costs, since one of the organizations has only 30 resident members, while the other has 121 resident members as well as several organizational members, the burden on the smaller organization has turned out to be too large, especially since the maintenance of the clock was assigned to the smaller organization. Furthermore, the rowhouses and live-work units that were sold are not part of either organization—a common practice in the Netherlands. In retrospect it seems as if having a single residents' association would have been a better way to structure the organization, and to consider including the house owners in the group. As now, the payments made by each owner would be prorated based on their use of the complex.

A third lesson relates to the public amenity of the clock. The clock tower was designed to serve the wider community of Barendrecht as a landmark and was paid for by the housing association, but its maintenance, which has turned out to be substantial, was assigned to the smaller resident association with thirty members. When including a larger community amenity in a project, the assignment of the long-term maintenance needs to be carefully thought through, assuring that the municipality or the housing corporation cooperate in the maintenance costs.

A fourth lesson derives from the lack of differentiation between lower-income and middle-income units. The uniformity of the typical unit design was likely due to its being an economical choice. But having the same design for both the social housing units and the middle-income units, no doubt contributed to the

The clock

marketability of the social housing units, and allows low- and moderate-income people to live side by side with no one being able to identify a person's income from the unit design.

A final lesson relates to the role of architectural design. In many ways, for the Dutch context, this project is conventional in the way that the units are designed and grouped. However, the complex is very well-organized to create visual variety, and designed with great attention to aesthetics and details. While brick is a common material in the Netherlands, the cheerful color enhances the north façade of the plaza, combined with the openings and massing that are highlighted with white panels and white concrete effectively reduce the scale and increase visual interest. The structure reads as a set of parts that gather to create a unified whole. By addressing such things as the proportions of the tower, and how to make a large façade facing the square human scale, the project is understated, elegant, and highly suitable as a city landmark that fits its context.

Notes

1. Roelf Steenhuis (Steenhuis Bukman, architect, Carnisselande), Interview, Delft, March 8, 2010, 2; and Roelf Steenhuis, Interview, Barendrecht, June 24, 2016, 1.
2. Gerard and Wil Onder de Linden (Residents, Carnisselande), Interview, Barendrecht, June 24, 2016, 1; and Koos and Gonny van der Elst (Residents, Carnisselande), Interview, Barendrecht, June 24, 2016, 2.
3. Marianne Laurens, "Wonen in een Vinexwijk," *Via Vestia*, no. 1: Rotterdam (Summer, 2011): 18.
4. Koos and Gonny van der Elst, Interview, 2016, 2.
5. Ibid., 1.
6. Roelf Steenhuis, Interview, 2010, 2.
7. Ibid., 6.
8. Roelf Steenhuis, Interview, 2010, 2.
9. Gerard and Wil Onder de Linden, Interview, 2016, 2–3; and Koos and Gonny van der Elst, Interview, 2016, 3.
10. Koos and Gonny van der Elst, Interview, 2016, 3.
11. Ibid., 3.
12. Roelf Steenhuis, Interview, 2016.
13. Koos and Gonny van der Elst, Interview, 2016, 1.
14. Ibid.
15. Ibid.
16. Ibid.
17. Gerard and Wil Onder de Linden, Interview, 2016, 1.
18. Ibid., 3; and Koos and Gonny van der Elst, Interview, 2016, 2.

6 De Beeklaan

BULLET 2
BULLET 2

De Beeklaan tower at the corner of Loosduinseweg and Beeklaan

Location of De Beeklaan in the Netherlands

Location of De Beeklaan in Den Haag

Location of De Beeklaan within its neighborhood

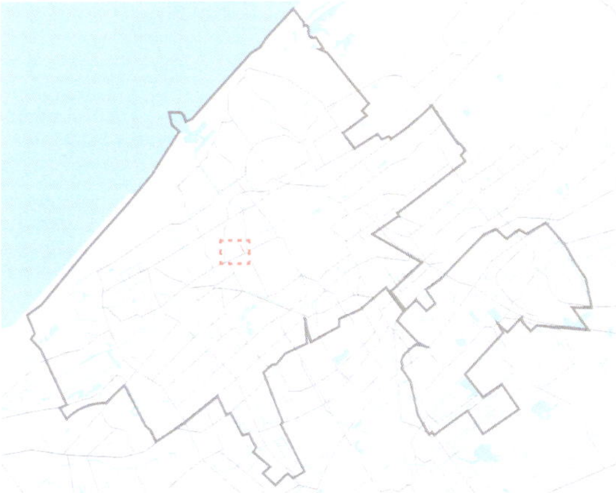

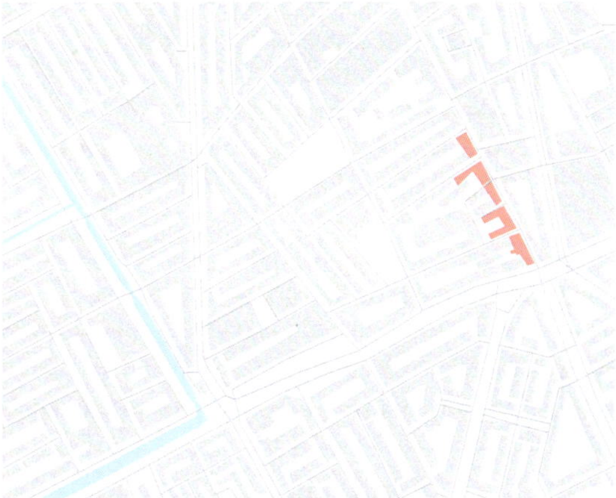

De Beeklaan is located in The Hague (Den Haag in Dutch). Designed by SCALA Architects and developed by Bouwfonds Property Development (BPD), it was completed in 2007. The project is selected as a representative of traditional design or a New Urbanism approach, an important architectural movement in the Netherlands.

With a density of 200 units per hectare (80 units per acre), the project follows the principle of housing as urban fabric by encompassing part of five small blocks in The Hague with housing designed to fit the existing pattern of rowhouses on the side streets and shops with flats above on the commercial street, although at a slightly higher density than existed. A tower on the southeast end signals its status as a landmark and entry to the neighborhood. Although largely social housing (56%), the project also includes middle- and upper-income units. These take form as gallery-accessed flats, rowhouses, penthouses, and a senior housing community alongside commercial space and parking.

De Beeklaan is named for the street it faces in the Regentessekwartier neighborhood, one identified as needing a radical intervention. The continuing presence of crime in the area drove the decision of city, the police, and the neighbors to demolish substantial portions of the five blocks and start fresh.[1]

A competition for redesigning the neighborhood was held and won by SCALA architects with developer Roey that had resulted in reconstruction of three streets of new housing next to the proposed site in 1999. The competition project, Woningen Regentes-Valkenbos, designed based upon the principles of 19th-century urban housing of Den Haag, used such characteristics as brick "ornamented with gables and decorative bands in the masonry," bay windows, and front gardens, while including

De Beeklaan Block I tower and the rest of the project along the street beyond

The neighborhood adjacent to De Beeklaan

21st-century amenities.[2] The Beeklaan site, however, was awarded by the city to another developer, BPD, that was planning to work with a different architect. But the neighborhood group, cherishing the 19th-century style of the remaining buildings and desiring to maintain the neighborhood's historic character, convinced the city and then the developer to work with SCALA as the architect.[3]

In response to the neighborhood interests, BPD agreed to work with SCALA to rebuild the neighborhood. When the original design came in way over budget, the city did not want to reduce the quality of the housing construction, so changes were made such as reducing the size of units,

simplifying the architectural details, and eliminating the balconies. Additionally, it was found that because of the vibrations from the nearby tram, the original plan to put parking underground would not work, and the car and bicycle parking had to be placed on grade. Finally, the developer, concerned about the street noise from De Beeklaan, added thick, sound reduction glass to the windows on the north side.[4]

This was the first urban project developed by BPD. Prior to this the company had developed very profitable projects outside the urban area—"easy building because there is nothing there."[5]As its first urban project, BPD was very concerned that it be successful and went out of its way to create a fine project.

The De Beeklaan project was conceived just prior to the recession and built as the recession developed. In the end, Bouwfonds considered the project was a success even though it lost money on it.

> We have succeeded because it is . . . a very beautiful place architecturally. You can discuss whether it worked. We hoped that this would generate improvement [in the overall neighborhood], but it hasn't yet. That this [project was actually built] is a gift to the city and to the neighborhood. . . . Mieke Bosse [of SCALA] did a great job.[6]
>
> We see it as a PR project. We earned nothing (actually we lost a lot of money), but because we earned a lot of money outside the city we could do this for the city.[7]

When asked what was the biggest challenge to getting the project built, project manager Wicher Mol said "The economic mix. Bouwfonds is a rich and large company. A small company never could have made this project."[8] There were problems with the contractors due to the recession. "The first one went bankrupt . . . so we had to find another in a few months, when a lot of building was going on. It was difficult . . . and it was expensive."[9] He also mentioned that it was difficult to sell the units.[10] This was in part due to the recession, but also according to the architect Mieke Bosse, BPD was inexperienced in selling housing in the urban setting.[11] BPD ended up selling the cooperative units for the elderly and some of the rowhouses at a reduced price to Vestia, a social housing corporation.[12]

De Beeklaan diagram showing parking, apartment blocks, and rowhouses

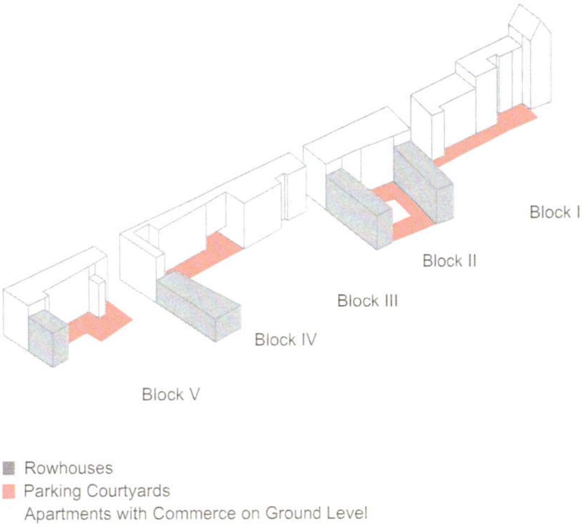

Site, Context, and Building Functions

Addressing the neighborhood group's desire to rebuild the area similar to what is there, the De Beeklaan project combines flats above the commercial space at ground level along the Beeklaan with rowhousing that lines the side streets. In order to meet both the desire to house the existing residents, predominantly renters, and to bring others to the area, the flats were mostly designed to rent, and the rowhousing was designed for purchase.[13]

De Beeklaan floor plans: levels 0 to 3

De Beeklaan floor plans:
level 4 to roof

De Beeklaan Block I northeast façade

De Beeklaan Blocks III and IV connect at level 1 to form the senior housing DOG

The tallest block, Block I, located at the corner of two main streets, Beeklaan and Loosduinseweg, has a tower that indicates the entrance to the neighborhood. It comprises commercial space at the prime corner on the ground level, with parking behind the Beeklaan façade, and gallery-accessed rental flats for middle- and upper-income residents on the five levels above. The parking area is entered from the adjacent side street, Noorderbeekstraat.

Block II has commercial space along the Beeklaan façade on the ground level with rental flats above for middle-income and social housing. Facing Noorderbeekstraat on one side and a little street also called Beeklaan on the other are twelve rowhouses (six on each street) that have front entrances on the street and parking at ground level under decks at the back. The parking for the flats is under the main building, and is entered from Noorderbeekstraat.

Blocks III and IV have been combined to create an L-shaped Groepswonen Door Ouderen (DOG) on the three levels above the street. A DOG is a community for people over age 50 who live independently, want their own front door and personal freedom, but also seek common space for activities with others[14]. Existing three-story structures at the back of the two blocks have been retained, while the ground-level spaces fronting Beeklaan serve commercial functions. Parking is located at ground level both behind the commercial space and behind the Teijlerstraat façade. In the middle of the two-block structure, the three levels of gallery-accessed DOG flats form a gate over the street (De Gheijnstraat). Along Beeklaan the building is four levels tall, while along Teijlerstraat, bordered by rowhouses, the building is only three stories in height and consists of two levels of flats with parking below.

At ground level Block V serves commercial functions along Beeklaan with parking behind. Above ground level are three levels of gallery-accessed rental flats serving middle income and social housing residents, and entering from the north on the side street, Vinkenstynstraat, are two rowhouses. The entire block is four stories tall. Existing rowhouse structures with gardens behind form the remainder of the large block.

De Beeklaan Block II southeast (with typical parking access) and northeast façades

Parking courtyard for Block V showing cars and bicycles under rowhouse decks ahead, and under the gallery to the right. Photo by Mieke Bosse, SCALA Architecten

Organization, Appearance, Form, and Massing

As might be expected from a traditional design based on 19th-century principles, De Beeklaan is massed as a series of city blocks, with main façades of dark red brick with white bands of masonry. Special attention is paid to the corners, where shop entrances and the windows above are on the bias, vertically expressed with French balconies, a slight reveal, and crenelated parapet.

At six stories tall, the tower block at the corner of Loosduinseweg and Beeklaan is the tallest. Along Beeklaan, the bottom four levels of Block I are articulated as unified, while the top two, due to the use of light colored brick and setbacks, are articulated as having an independent identity, thus perceived as lower in scale from that side (see photo at left top of page 138).

Along Beeklaan, the commercial space is uniformly expressed with a white band across the lowest level and large windows that contrasts with the darker brick above. The horizontal brick façade is broken up with vertically articulated areas that have wooden bay windows topped with balconies and a crenelated parapet, and horizontal areas with lighter brick on the top level, and everywhere with decorative brick patterns.

Similarly, the four-story rowhousing on Noorderbeekstraat is mostly of dark brick (see bottom photo on page 140). The middle units have bay windows supporting balconies variously at ground level or the level above, and at the top level a sloped roof with dormers. Some of the entrances to the middle units have a tiny garden space by the front door. At the end of the rowhouse blocks the last house is vertically oriented with a full-height four-story masonry tower that is fenestrated with French doors and Juliet balconies.

On the three-story Teijlerstraat façade, the ground level parking area is faced with large window and gate openings that are covered with cast iron grills typical of a garden fence, which softens the appearance at the sidewalk. The two upper levels of housing incorporate bay windows and light brick at the top that reduce the apparent size of the façade. The overall effect blends with the 19th-century adjacent buildings to create a lively, human scale streetscape.

The commercial level is articulated as a white band: Block V

The rowhouses on Noorderbeekstraat show full-height towers, bay windows, and pitched roofs with dormers (southwest and southeast façades)

An outdoor living area is created along the extra-wide southwest-facing gallery on Block V

Unit Typology

The project goal was to provide housing that would fit in the urban context and serve people with high, moderate and low incomes. Because the noisy street is to the northeast, it was important for the units to have access to the southwest. Therefore, most of the units are gallery-accessed flats, about one-third of which form the assisted living housing community for the elderly. The remainder is largely for middle- and upper-income residents. The fourteen rowhouse units are also designed for middle- and upper-income residents. Several of the flats in Block I are penthouses, designed for the luxury market. Although full-scale balconies were not possible in most units due to the budget, many have Juliet balconies (where a full height, operable window is covered by a railing to prevent falling out) that allow interior spaces to have access to the outdoors.

Gallery Flat

The most commonly used unit in the project, including the assisted living, is a two-bedroom gallery-accessed flat that is found in the center of the blocks. These units have a bedroom and living room with windows

De Beeklaan unit plans: A

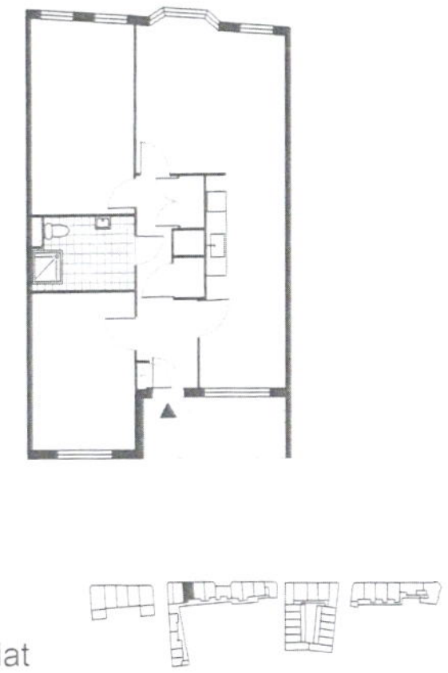

a. Gallery Flat

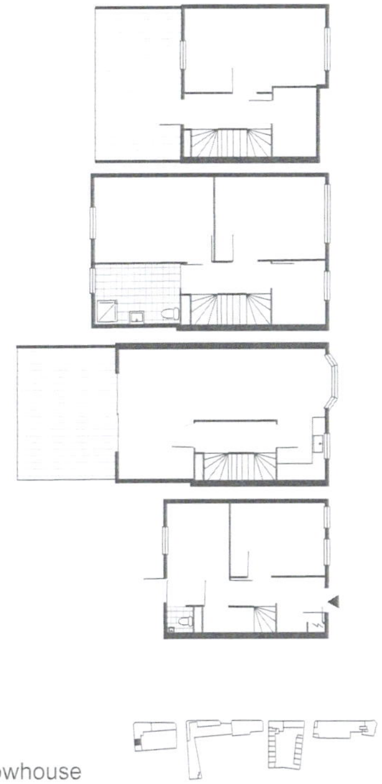

b. Rowhouse

0m 2m 5m

1:250

The living room and deck of the Vinkenstynstraat rowhouse

facing Beeklaan and another other bedroom on the gallery side. The kitchen and dining area also face the gallery. While there is no balcony or loggia due to the budget constraints, the southwest-facing galleries are designed to be extra wide to accommodate outdoor living in the part of the gallery next to the unit entrance that is protected by the projecting second bedroom. One enters into a typical Dutch vestibule corridor with doors to coat closet, toilet, bathroom, bedrooms, dining-kitchen, and living room, the latter two spaces being interconnected. In this design the living room includes a bay window overlooking Beeklaan rather than the French doors and a balcony found on other levels.

The decks for the rowhouses on Vinkenstynstraat and galleries of Block V with parking courtyard below

The penthouse unit on the top floor of the Block I tower

Rowhouse

The rowhouses along Vinkenstynstraat are three stories in height with auto and bicycle parking at the back underneath the living room deck (see top photo previous page and plan b, page 141). Each has two ground level entrances. The first entry comes directly from the street into a corridor that contains the staircase and connects to the two ground floor rooms, one at the front of the house that could serve as a bedroom. The second entrance, from the rear parking area, enters directly into the second small room, for storage, sleeping or study, that connects to a toilet and under stair closet. Such a traditional controlling corridor is located on each level of the house.

The first level incorporates the living area, dining area, and kitchen in one large space. It opens onto the southwest-facing deck over the parking spaces and overlooks the parking courtyard and gardens beyond (see bottom photo previous page). By the dining area is a bay window with a view to the street.

De Beeklaan unit plans: B

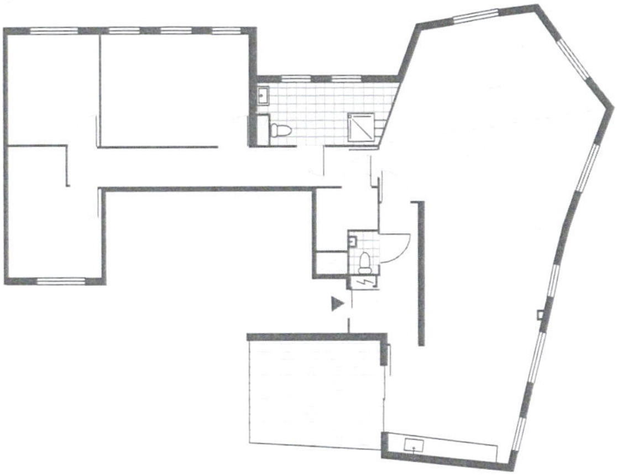

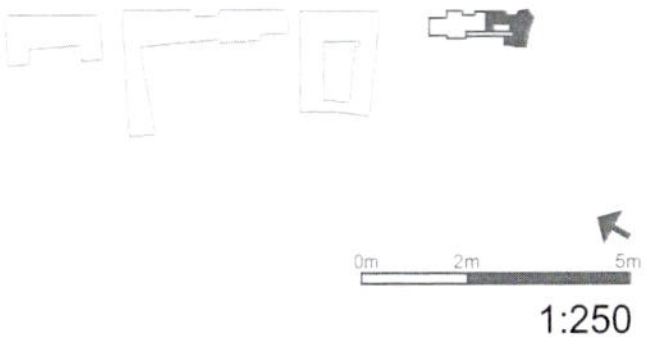

c. Penthouse

De Beeklaan space syntax gamma diagram

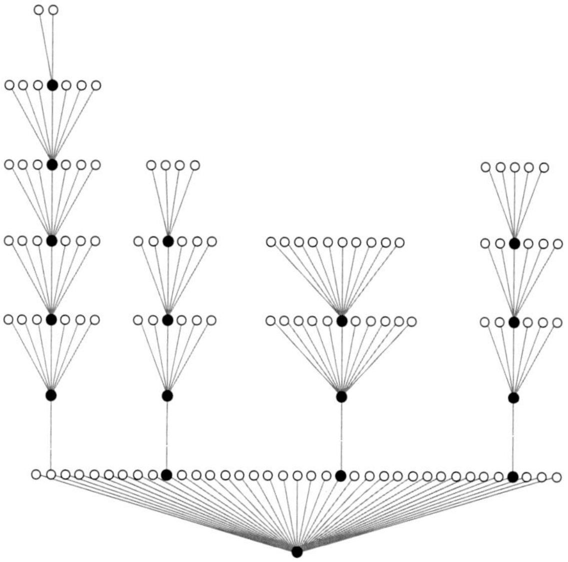

On the second level are two good-sized bedrooms, a full bath, and a storage room. In the front bedroom, French doors open onto a tiny balcony over the bay window. The back bedroom looks over the parking courtyard. The third level consists of a third large bedroom along with a laundry room and roof deck accessible from the stair landing. This bedroom has French doors and a Juliet balcony.

Penthouse

This penthouse unit is on the top floor of the tower building, with a very large living room beneath the tower, reflecting its round shape, extra height and open to dining area and kitchen (see plan and photo page 143). The corridor arrangement is not typical, being a series of corridors that require passage through active spaces. Once on that floor from the elevator vestibule you enter another hall that contains the toilet as well as doors toward the kitchen and the living areas. This dramatic living space has six windows along both the northeast and the southeast sides, four of which are French doors with Juliet balconies. At the kitchen area sliding doors open onto a southwest-facing roof deck. The enclosed corridor to the private rooms is accessible from the living room, and connects to the three bedrooms and full bath.

Syntactical Structure

De Beeklaan's gamma analysis diagram is very direct and simple. Each block has its own independent identity with commercial and rowhouses entering at the lowest level directly from the street, and flats along corridors above the ground level. Some blocks have more units; others have fewer. The Block IV rowhouses are in a separate thread from their associated flats as they are on a different street.

Social Principles: Collectivity, Income Mix, Tenure, and Social Interaction

De Beeklaan was designed as part of a neighborhood rehabilitation effort in which the local residents were involved. As the neighborhood had a large proportion of elderly, Blocks III and IV were designed to form a cooperative housing project for the elderly. There was a desire for more of a mix in income, so a mix of income levels and unit types was provided, but with more than half serving as social housing, especially for the elderly in the cooperative housing in Blocks III and IV. As stated earlier, the rowhousing was intended to be purchased, providing more stability to the neighborhood. Each block houses no more than two different income levels.

The limited number of flats on each gallery assures that neighbors will recognize each other, with the extra-wide

Neighbors gather in the Vinkenstynstraat next to Block V

size providing space for outdoor activities and social interaction. The use of rowhouses along the side streets creates activity on the street and a great potential for neighborliness, as was observed in adjacent buildings when visiting the site.

Lessons and Conclusions

Although a traditional approach may limit the formal options for complex housing, it does not at all rule out using the principles to create architecture for dense housing. De Beeklaan shows that a traditional approach can effectively complement complex housing principles. In fact, some principles like mixing housing with other functions are implicit in traditional approaches.

Having a traditional design contributed to the project's success. The developer was very content with that approach. "We love the historical style. In the Netherlands you can call it a very commercial style because the people who are buying love that style."[15] Furthermore, the residents wanted the traditional design. Being in a low-income neighborhood made the project especially challenging, even though the neighbors played an important role in selecting the traditional design.

Architectural detail of De Beeklaan. Photo by Mieke Bosse, SCALA Architecten

But the replicability of the approach taken at De Beeklaan to other contexts may be limited, as the developer was willing to take more risk because the construction of the first urban project served a public relations purpose.

Perhaps the most important lesson from De Beeklaan is the necessity of having a developer who is committed to

making the project work, and the advantage of working with a large development company. The recession made achieving this project especially difficult, increasing the challenges of getting it built at a reasonable cost, and in the end creating difficulty in selling the units.
Nevertheless, the developer was motivated to see the project through, had enough economic power to do so, and was happy with the result.

Notes

1. Mieke Bosse (architect, De Beeklaan), Interview, Den Haag, May 22, 2012, 1.
2. SCALA website, Woningen Regent Valkenbos, Den Haag, November 1, 2015, www.scala-architecten.nl/project/regentes-valkenbos.
3. Ibid., 2–3.
4. Ibid., 3.
5. Wicher Mol (Project Manager, Bouwfonds, De Beeklaan), Interview, Den Haag, May 12, 2012, 1.
6. Ibid., 1.
7. Ibid., 1.
8. Ibid., 3.
9. Ibid., 3.
10. Ibid., 3.
11. Mieke Bosse, Interview, 2012, 3.
12. Wicher Mol, Interview, 2012, 3.
13. Mieke Bosse, Interview, 2012, 5.
14. Groepswonen Door Ouderen. http://www.gdo-denhaag.nl/.
15. Wicher Mol, Interview, 2012, 1.

7 De Opgang

De Opgang: southeast corner on the Tussenmeer

Location of De Opgang in the Netherlands

Location of De Opgang in Amsterdam

Location of De Opgang within the neighborhood of Osdorp

In 2005, an Amsterdam city alderman, Simon Willing, approached private developer Henk Jan Hollander with a question. Willing wanted to save the last active church in his district and wondered if the developer had any suggestions as to how this might be done?

The existing church building was much too large for the congregation, very expensive to maintain, and the members were having difficulty supporting the salary of the minister. Hollander proposed using the church site and an adjacent city-owned lot to create housing that the city, the church group, and the developer would jointly support and jointly profit from. The goal was that the church group's profit from the sale of the housing, when invested, would be sufficient to support a minister's salary as well as the maintenance of the church and its activities.[1]

Working closely with the church community, Hollander's first step was to gain the members' confidence, for in the Netherlands developers do not have the best reputation for serving people's needs. An unusual developer because he was trained as an architect, Hollander was committed to having an outstanding architectural design firm design the project. To help the community feel comfortable with the design process, as a part of the discussion of the program for the building Hollander asked a colleague to make a stylebook of eleven or twelve different styles of church, each associated with a list of architects. "I showed them and I said, 'You can choose.' "[2] Although he expected the congregation to be very conservative in its choice, in the end it selected a contemporary style associated with the firm KCAP Architects&Planners, whom the congregation chose as their architect.

KCAP's first action was to create a series of ten or more alternative designs that responded to the programmatic concerns that the congregation had developed with Hollander.[3] For example, the church had to be visible from the street so that it had a clear identity; it had to

View of De Opgang sanctuary from the Hoekenesgracht Canal ©Pauline Borst

remain a landmark; and the housing could not be too high or there would be problems from the neighborhood. The most important decision was whether or not to keep the existing church building, so KCAP's schemes included both options. Both Hollander and architect Han van den Born found the church members committed to the best solution: "The church people were very open-minded."[4] While not an outstanding example of church architecture, the original building had some historical value, but it was designed for 1,000 people and the present congregation was 100. Furthermore, it was an introverted design, and the congregation wanted a building that would invite community participation. The architects and the developer, having mixed feelings themselves, realized that any decision to demolish the existing building had to come from the congregation. The congregants chose to build a new building.

One Sunday, after the church service, KCAP was asked to present the new design.

> We came up with a simple idea. The idea of a pearl in the oyster shell. The pearl in the middle is the church. On the outside it is quite rough. Make a world of wood like a big roof. The church is small, but the housing on top makes a

shelter to define the entrance area. . . . Along the green park is a window toward the more quiet direction—the green and the water. A lot of people are driving on the road, they can still have a view of this big window. At night with the light inside, the cross is visible.[5]

Site, Context, and Building Functions

Located in the neighborhood of Amsterdam Nieuw-West (formerly Osdorp), De Opgang (The Ascension) comprises eighty dwellings, a church, a childcare center, a community meeting space, and a parking garage.[6] The housing is 240 units per hectare (100 units per acre). With a canal and greenway to the east, and a large thoroughfare with tram stop as well as a park to the south, the project is centrally located with access to plenty of green space.

The church serves as a community center, with a large dining area that accommodates public meetings and also several smaller meeting rooms.[7] The sanctuary, with its large stained glass window, natural wood walls, and tile floors includes flexible seating.

As developer, Hollander had identified an unmet need in the neighborhood for housing for single people and young couples, so the eighty housing units are mostly one- and two-bedroom flats with a few family rowhouses on the ground level. The church program also included a childcare center that enters from the main street, Tussenmeer, and community space

De Opgang viewed from the southeast ©Pauline Borst

Sanctuary of De Opgang church
©Pauline Borst

Meeting and dining area of the church

Northwest corner: rowhouses with flats above

Garage entrance with view to the courtyards

associated with the church. And despite the fact that many people in the neighborhood do not own cars, the city required a parking place for every unit, so the project incorporates a garage.[8] Problems with the geology dictated that the parking ramp be partially above ground.[9] As a result, the church is placed on a plinth above the street level with its entry protected from rainy weather by a large cantilevered roof formed by housing above. The apartments are arranged to create three courtyards, two above the church for the residents, and another at the level of the church that serves the childcare center.

De Opgang floor plans

Organization, Appearance, Form, and Massing

De Opgang is a layered building with parking at the bottom and access to the residences via two entrance lobbies on the east and west, with elevator shafts and staircases. The parking level is one half-level underground. The next layer, or lower level, includes the ramp to the parking, and at the west end has rowhouses and access to the west entrance lobby and bicycle parking. On the south side there is street level access to the childcare center and a half-level up access to the church. At the northeast corner again a half-level up

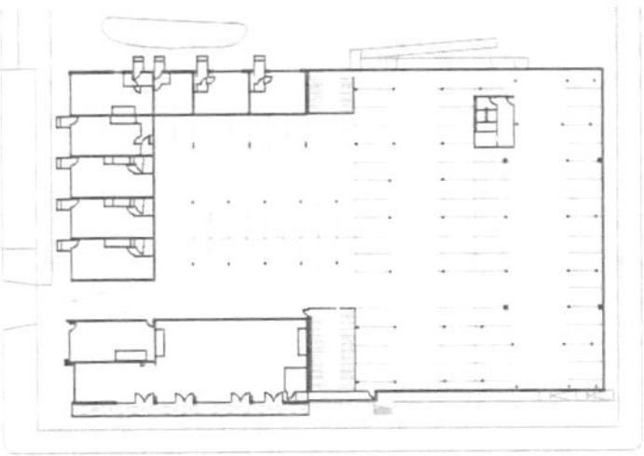

Parking Level

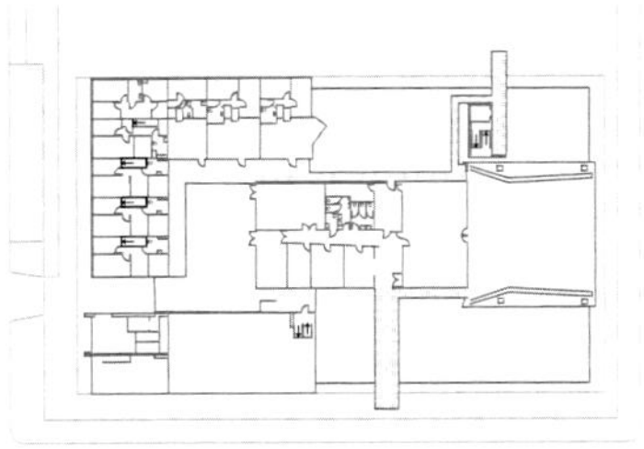

Ground Level

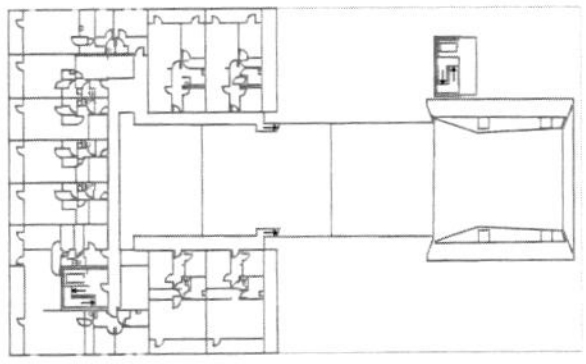

2nd Floor

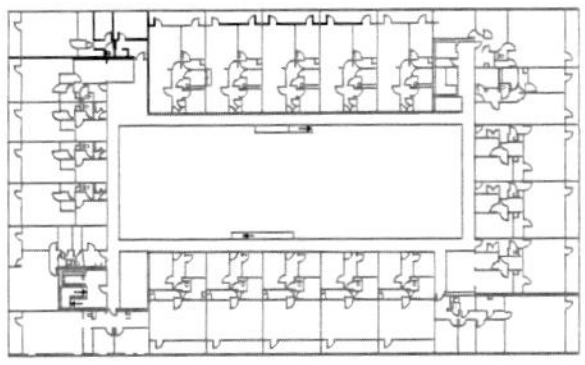

3rd Floor

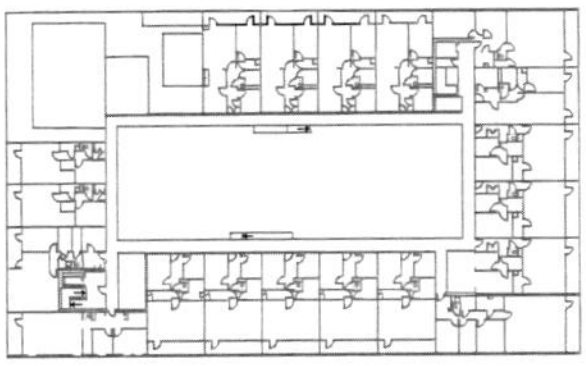

4th Floor

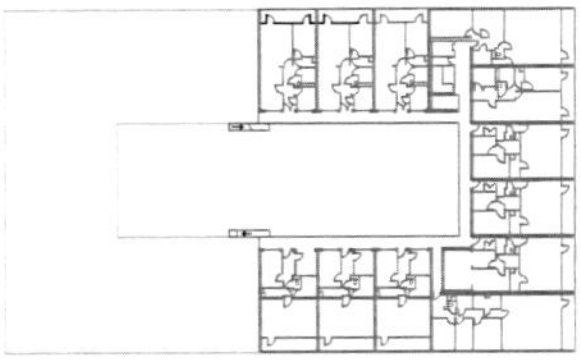

5th Floor

6th Floor

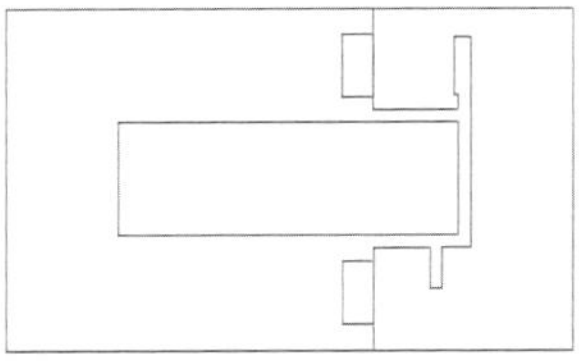

Roof Level

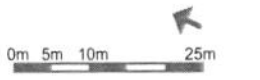

1:1500

De Opgang building section

De Opgang circulation diagram

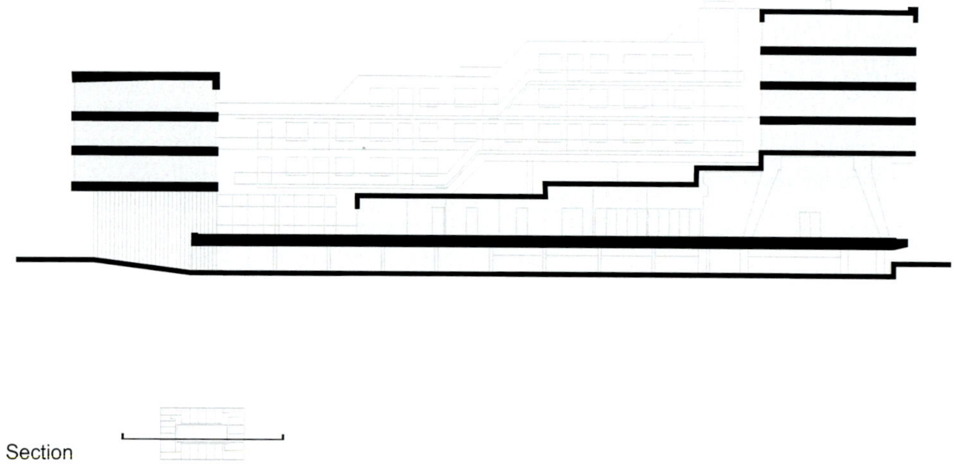

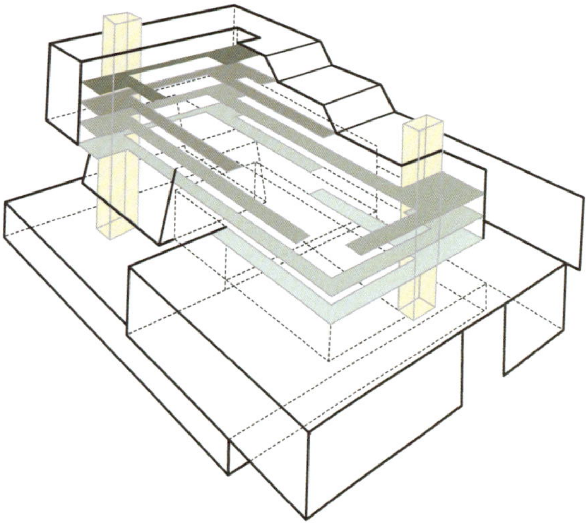

from the street is the east entrance lobby. While the rowhouses have mail delivery directly to their door, each entrance lobby contains the mailboxes for approximately half of the apartments. The developer realized that the busy street was a good place to access the church and the housing, and he wanted activity on the street.

> Another key element was the ground level. There should be as much life as possible. That is the reason for the childcare center, and the maisonettes that give you a lot of entrances on the ground level. This gives a lot of life, a lot of social control.[10]

The main level, here called level 1, comprises three parts: the church and its associated community spaces at the east end, the childcare center on the south with the children's courtyard in the middle, and the top layer of rowhouses, two of which, on the northwest corner, have terraces adjacent to the children's courtyard. Level 2 exists only at the west end of the site. It consists entirely of apartments that are accessed through a U-shaped gallery that overlooks the children's courtyard.

Level 3 is the only level with a full complement of twenty-four apartments. On this level, the dwellings form the edge of the building, and are entered through the gallery that encircles both the children's courtyard and the residential lightwell courtyards. The gallery is reached from both the east and west elevator cores, thus providing access to gallery levels for all flats.

On the subsequent levels the building, full height on the east side, steps back from the west end as the number of apartments is reduced (twenty-one on level 4, twelve on level 5) until level 6, where there are only eight apartments along the east end. This graduated scheme

Play courtyard for childcare center

Southwest corner: childcare façade and access and church recessed entry, all on the busy Tussenmeer street

Galleries and courtyards looking east

East façade showing the contrast between the green tiles and the wood around the sanctuary

allows for light to penetrate into the courtyards, and reduces the apparent mass of the building. For levels with units only reached from one elevator core, exit stairs to the next lower level gallery are provided within the courtyard.

The architects were very concerned that the church have an identifiable and distinct expression independent of the housing. In addition to providing the large stained glass window as a beacon on the east, the design raises the church on a plinth, and places the housing as a block raised on columns that protectively hovers over the church like a hen above her eggs. The building setback with its contrasting materials of the wood and green tiles around the church and green brick facing and clear glass balconies for the housing reinforce the independence of the two functions. On the west end the expression of the housing block raised above the church is maintained by continuing the cantilever of the apartments above, by having vertical wood bearing walls below and by facing the rowhousing with wood and glass in contrast to the green brick (see top photo opposite page). A similar approach is taken on the south façade, with the childcare center again faced with vertical wood elements and glass. The negative space of the church entrance is reinforced by the wood- and glass-covered wall of the childcare center in contrast to the green brick facing on the housing above. At the same time, the housing block appears to be less weighty because of the use of balcony openings that create cavities on the east and west and form a screen layer along the north and south walls.

West and south façades showing material differentiation and balcony recess on the west

North façade: screen layering and step-back of the balcony wall

Housing and Unit Typology

This project was designed with a variety of unit typologies in mind. Hollander, the developer, wanted "80 dwellings divided into a lot of different floor schemes."[11] All units were to be sold and the building was to have a cooperative governance. Hollander was very clear about the key issues to be addressed by architects in the unit design.

> The apartments on the south side have the disadvantage of the traffic noise, but the advantage of the park view and the sun. The north side is the quiet side. During the selling of the dwellings there was a 50/50 division between the dwellers. It was really balanced. For me it is always a big compliment when I have predicted the market. The apartments on the north side have a balcony to the northwest, with the living and kitchen space open from one side of façade to the other. Open through to both sides. The south side apartments have a balcony, living and kitchen on the south side with sleeping rooms on the gallery. Some people hate to sleep on the gallery . . . and in the corner you get some special apartments 3–4 rooms . . . [also] maisonettes.
>
> [I wanted to provide] large balconies, [also] having toilets separate from bathrooms (when you as a visitor go to the toilet, it is embarrassing to see all the private particulars). Storage in the apartment is important. And living rooms as big as possible, [with] the possibility to separate the kitchen from the living rooms in certain apartments (north and corner) 50% like to combine with living. Turkish people like to have a separate kitchen. About 50% of the residents have a Dutch background and 50% have non-Dutch background. . . . What is very important for me is that you can open your windows. I like natural ventilation as much as possible.[12]

In the end, there were eighty units of six basic unit types (we show four of the six unit plans below) designed as described earlier according to the placement in relation to context and to location within the building:

1. Rowhouses at the base (two types)
2. West side one-bedroom apartments
3. West corner two-bedroom apartments
4. South and east side two-bedroom apartments
5. East corner three-bedroom apartments
6. Northside two-bedroom apartments.

De Opgang unit plans

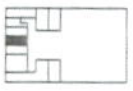

a. Rowhouse

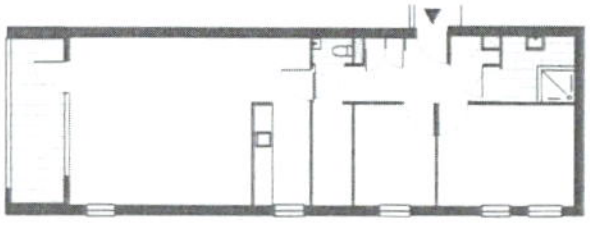
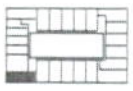

b. West-Facing Corner Flat

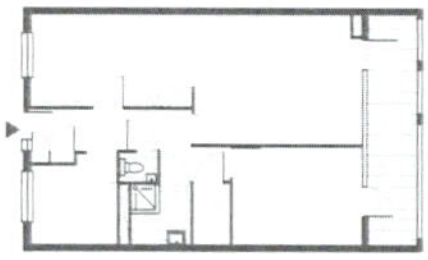
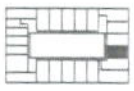

c. East and South-Facing Flat

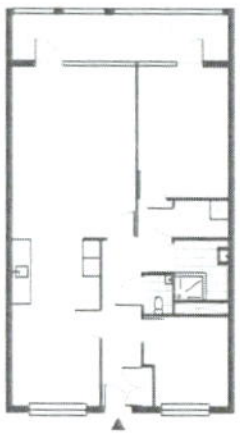
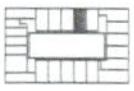

d. North-Facing Flat

The rowhouses have direct access from the street, while the apartments are all reached through the galleries.[13]

Entry to a west-facing rowhouse

Rowhouse

The eight 2-story rowhouses are entered through terraces that are slightly raised up from the level of the side street (see photo below), Wolbrantskerkweg, and therefore removed from the heavy traffic along Tussenmeer. The wider front terraces serve as outdoor space for the five units that face west. The ground-level living room of these typical units has light and ventilation from the west side, while the two-story kitchen space is open to a clerestory window from the children's courtyard on the level above (see plan a to the left). The upper level has two west-facing bedrooms as well as toilet and bath. The north-facing rowhouses are organized differently with entry, kitchen, and dining on the ground floor and bedrooms and living room on the second. These units have a south-facing terrace overlooking the play court on their upper level.

East-Facing Flat

The eastern, gallery side of these basic apartments—designed as social housing—includes a good-sized

bedroom and entry that face the gallery. The living rooms of the thirteen units include a kitchen wall and look west onto the side street, opening to both a balcony and a loggia. Due to the step-back design of the building, one of the units on the fourth level has a roof terrace.

West-Facing Corner Flat

These five 2-bedroom corner units, also designed for rent as low-income dwellings, have the advantage of having no rooms with windows on the gallery (see plan b page 158). The living room with separate kitchen faces west and the two small bedrooms face north or south, depending upon the corner location.

East- and South-Facing Flat

The vast majority of the approximately thirty non-corner units in the south and east façades have two small bedrooms that are both along the gallery; a few have only one bedroom (see plan c page 158). The typical unit has a living room and a small independent kitchen that can connect to it. The difference between the south and east units is that the south units have sound-protective loggias because they face Tussenmeer, the main artery, whereas the east units that overlook the canal and park have both loggias and balconies. The step-back design provides roof terraces for two of the south-facing units on levels 5 and 6.

East-Facing Corner Flat

Wider than the west-facing corner flats, these eight dwellings have three bedrooms, two of which are good-sized and face north or south according to the corner location, with the smaller bedroom on the gallery. These generous units have the largest living rooms in the building, with a kitchen that could easily be separated from the living room. A balcony and loggia overlook the canal and park.

De Opgang space syntax gamma diagram

North-Facing Flat

Unlike the other flats, these fourteen units have defined living room, kitchen, and eating areas (see plan d page 158). The eating area overlooks the gallery, the kitchen forms the middle interior space, and the living room faces north. Of the two bedrooms, the smaller has windows on the gallery, while the larger one is oriented to the north. The living room opens onto a loggia, also to the north. Three of the north-facing flats are provided a roof garden by the step-back arrangement of the complex.

Organization and Syntactical Structure

The organizational structure of De Opgang is relatively simple. At the ground level, there are rowhouses, a commercial childcare center, and church that enter directly from the street, thus generating activity on the sidewalk. Above the street level, the building is structured around two elevator and stair towers, with the dwellings accessible from either or both towers, depending upon the level. Levels 2, 5, and 6 have only access from one tower; levels 3 and 4 have access from both towers. On levels 3 and 4, where both elevator

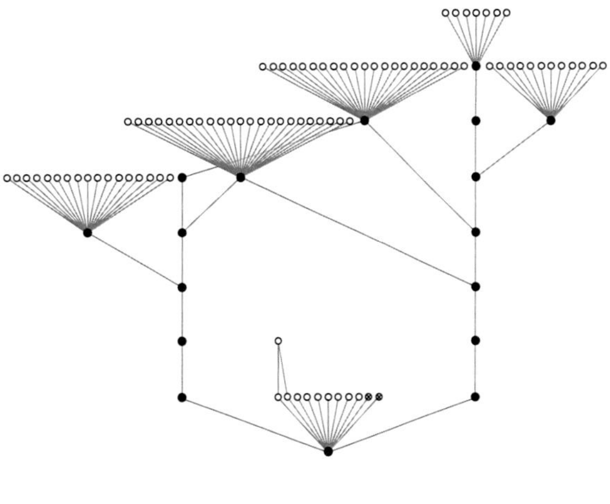

cores are connected via the gallery, twenty-five and twenty units are accessed from the gallery, respectively.

The gamma diagram shows one fan at the base and three threads, two of which are highly interconnected. The church, the childcare spaces, and the rowhouses form a fan at the base, all being entered directly from the street. The two threads that represent the two elevator cores link all of the apartments at levels 3 and 4. As shown in the fan-shaped patterns on the diagram, the galleries at each level collect and control access to all the dwellings at that level, ranging from the eight apartments at the top floor to the twenty-three and twenty-four apartments on levels 3 and 4. Although all seventy-two flats are interconnected, and levels 3 and 4 have a large number of flats, because there are two elevators, it is likely that people will tend to use the one closest to their unit. Therefore, they will actually encounter a relatively small number of neighbors on a daily basis—thirty-six units per elevator, and twelve or so units on each level. The syntax diagram shows a high level of internal connection, but little connection to the exterior.

Social Principles

The only required connection between the residents and the church is that owners of housing at De Opgang must sign a document acknowledging that they accept the ringing of the church bells on Sunday. But, as is typical with such projects, upon purchase of a unit, residents automatically become members of the cooperative that organizes the maintenance of the building and includes representatives of the church and childcare center. While a cooperative of owners forms an economic community, and renters are represented by the developer, such a community tends to support social networks among the building occupants as well.

Additionally, all of the flats in the building are organized in a shared circulation system that serves as a kind of connective tissue between occupants. The apartments are arranged at the perimeter of the building with the lightwell and childcare courtyards at the center. As the gallery and flat entrances are visible to all of the community members, inhabitants become familiar with each other by sight, if not with them personally, and are able to easily identify visitors or those who do not belong. A community terrace on level 4 is accessible to all of the flats.

Community terrace

Although eighty units is a large number, the organization of the units provides that nine of the units enter from the street and about thirty-six will regularly enter through each of the two entrances, thus breaking the overall number of residences in the project down into a more manageable number for each entrance.

The housing in this project, being all market-driven, has a smaller economic range than most of the other projects in this book. However, the combination of rented and owned units and the different unit types support a range of lifestyles, from townhouses for families that are connected to the street, to large corner apartments that might serve a couple with children, to small one- and two-bedroom flats aimed at young single people and couples.

Galleries and play court
© Pauline Borst

Lessons and Conclusions

De Opgang, still operated by the developer who built it, is remarkable due to its relation to its church community. The development, sale, and rental of housing on the site of a former church building provided the wherewithal for the church group to continue its existence and contribute to its community more effectively. Interest on the capital realized from the sale of the housing (the housing is owned by the developer and the remaining areas are owned by the church) is sufficient to fund a minister and church activities.

Essential to this process were an alderman who took an interest in preserving the church activities, a developer who helped to create a viable situation with such things as leases that give the church the right to ring its bell on Sunday and a church group that was open-minded enough to consider a very different type of facility. Adding to the positive mix is the location of the church across the street from one park and adjacent to another park alongside a canal. This provides green views on two sides, and because the housing is raised up, views over the adjacent buildings on the other two.

While the city played a role in generating the project and assuring sufficient land, unlike all of the other projects in this study, because there was no social housing included, the project financing was purely market-based, an investment by a developer willing to take the time to work with the church group.

This project not only shows the potential for development of housing on the sites of urban churches, but also shows how housing can be compatibly combined with them. As a result of the development, the church can continue its important community role.

Notes

1. Henk Jan Hollander (NL Development BV, developer), Interview, The Hague, May 23, 2012, 1.
2. Ibid., 1.
3. Ibid., 1.
4. Han van den Born (KCAP Architects&Planners, architect), Interview, Rotterdam, May 2012, 3.
5. Ibid., 3.
6. Henk Jan Hollander, Interview, 2012,3.
7. Gemeente Amsterdam, "Gebouwd in Amsterdam: De Opgang," ND, www.gebouwdin.amsterdam.nl/main.
8. Henk Jan Hollander, Interview, 2012, 5.

9. Ibid., 5.
10. Ibid., 3.
11. Ibid., 7.
12. Ibid., 7.
13. Hoekstra en van Eck NL, *Project De Opgang: Betaalbaar wonen in een bijzonder project in Osdorp.* [Real estate brochure] (Amsterdam: Hoekstra en Van Eck Makelaars, 2008).

8 La Grande Cour

La Grande Cour: east façade on the Westerdoksdijk

Location of La Grande Cour in the Netherlands

Location of La Grande Cour in Amsterdam

Location of La Grande Cour in the West Docklands

La Grande Cour was a central part of Amsterdam's urban development plan for the Westerdok or West Docklands. Close proximity to the train station and location on the harbor with views of the IJ River make the Westerdok a prime location. Arranged around three courtyards, La Grande Cour was designed by three different architects, MVSA Architects (formerly MVSA Meyer en Van Schooten Architects), de Architekten Cie, and Heren 5 Architecten. The project comprises 250 dwellings at the density of 350 units per hectare (140 units per acre), including the 30% of social housing required by the urban plan.[1] According to Rein Willems, the Bouwfonds developer (now Raboburt), the project was one of the first contemporary dense housing projects in Amsterdam that incorporated housing for family living. Because there was little experience with the mix of social housing, middle-income housing and high-end housing, the design "was an experiment," responding to the city of Amsterdam's commitment to maintaining an "undivided city" in which the rich and poor live side by side.[2,3]

The challenge was to create high density with a program that included units for purchase and rent and for three income levels (80 social units, 120 middle-income units and 50 high-end units). In addition to the housing, the project provides commercial space and a below-grade parking garage. At the time the project was begun in 1999, there was limited marketing research, so the developers had to envision for themselves who they thought would be the inhabitants for the middle- and high-cost units. Despite the tight housing market they knew many people wanted to live in this location, and that high-end purchasers would seek two parking spaces and good security.[4]

Although Amsterdam is generally a very strong market, in 2001 when the developer had to sell 50%–70% of the units to acquire the capital needed for construction, the market bubble had just burst. As a result, few direct buyers were interested, and Bouwfonds sold the middle-income units for less than top price to Deltalloyd, a pension fund. Thus the developers had less money than anticipated for construction.[5]

South façade cantilever and balconies

The startling building appearance of La Grande Cour, with the penthouse projections and balconies that make it a signature building, is primarily the result of an error in calculation. In response to the perimeter block urban design, Bouwfonds originally proposed the scheme as a three-courtyard scheme surrounded by eight stories of housing and divided into courtyards by two-story back-to-back rowhousing. But when the architect drew the first plans, it was discovered that the 250 units required by the city did not fit within the proposed building envelope as had been thought.[6] The project was about 50 units short.

Furthermore, the city did not want to make up the difference by simply adding more levels, as this would make the building appear too large and would cut off too much light to the adjacent areas. Instead, it was agreed that the designers would create the bent, L-shaped or periscope towers that have entrances adjacent to the rowhousing and cantilever over the other units at floors ten and twelve. The towers have a small footprint for seven stories (with one to three cluster apartments) and at the roof level of the original design, a large floor plate of two- and three-story volumes that cover a part of the existing roof area—and in two of the three cases, project over the street.

Additionally, the city, having experienced difficulty selling housing without exterior space, approved balconies for units that had been designed without loggias. Although the architects would have preferred a cleaner line with only loggias, the inclusion of balconies on the east and south façades and within the courtyards adds a play of shapes and shadow lines that enlivens the building appearance.[7]

Site, Context, and Building Functions

The Westerdokseiland area where La Grande Cour is located had begun to be developed several years earlier with the Silodam project. The island is located within the IJ River, and separated from the main city by a canal along which lies the Westerdok or West Quay. The urban plan, proposed by Peter Defesche of OD205 Architects and Dick van Gemeren and Jarne Mastenbroekof Dutch Architect studio, positioned buildings with six to ten stories on the island along the river, in contrast to lower, mostly three- to five-story development in the area on the south side of the canal.[8]

Located on the quay, La Grande Cour is a part of the second phase of development that includes three other complexes in adjacent sites, similarly predominantly housing with commercial space at the base and underground parking: the VOC Cour, Westerkaap I, and Westerkaap II.[9]

The West Dockland context

La Grande Cour diagram of parts

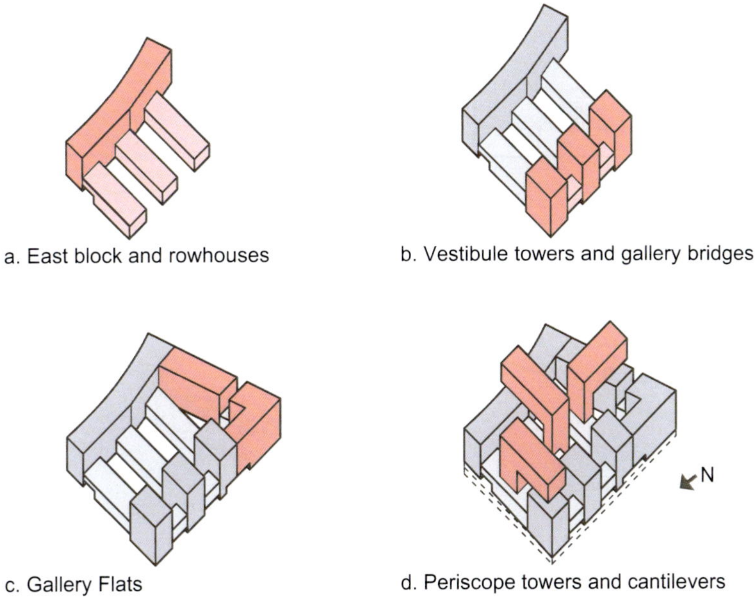

La Grande Cour floor plans: parking level to level 2

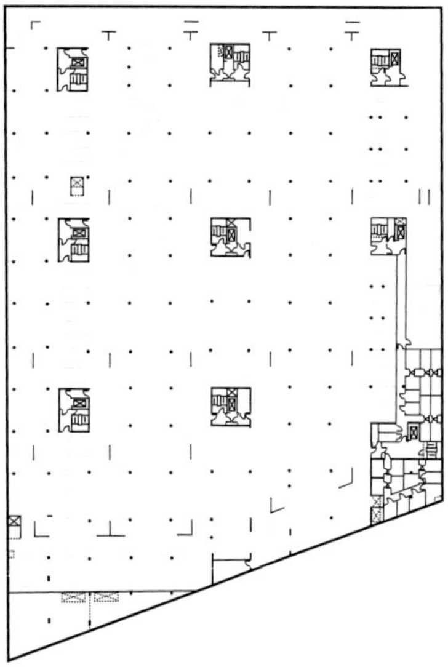

Parking Level

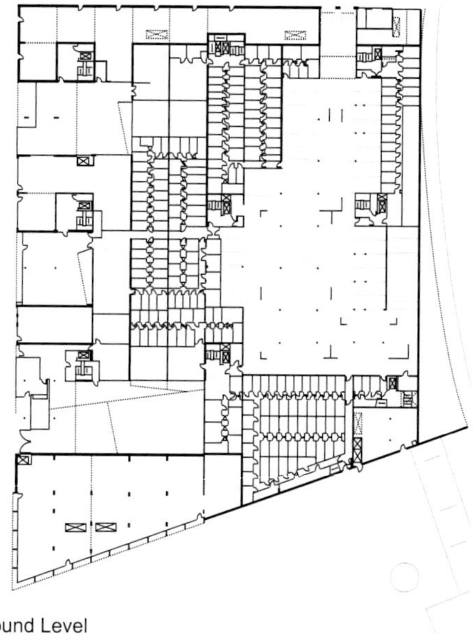

Ground Level

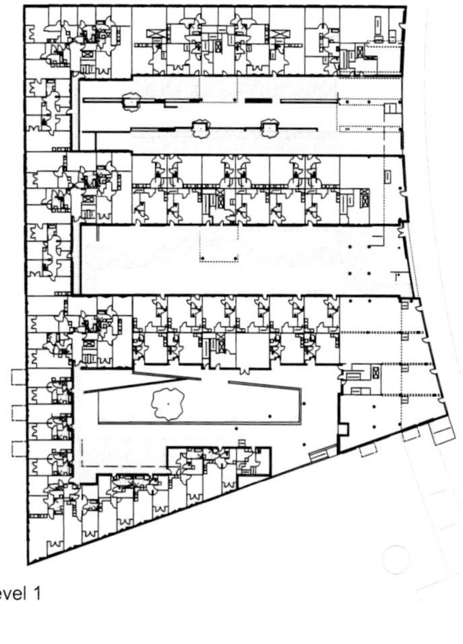

Level 1

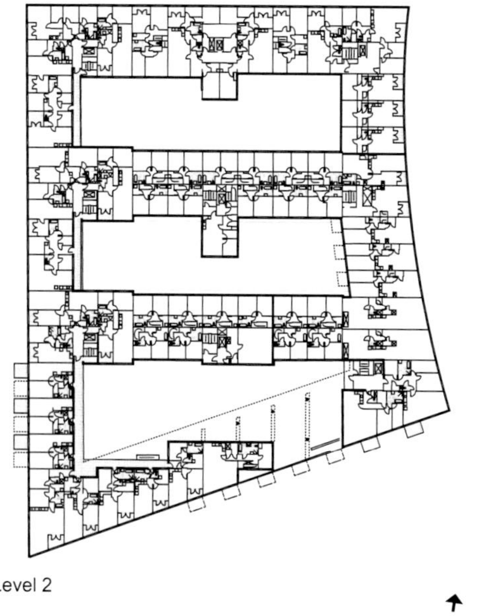

Level 2

0m 5m 10m 25m

1:1500

La Grande Cour floor plans: levels 3 to 6

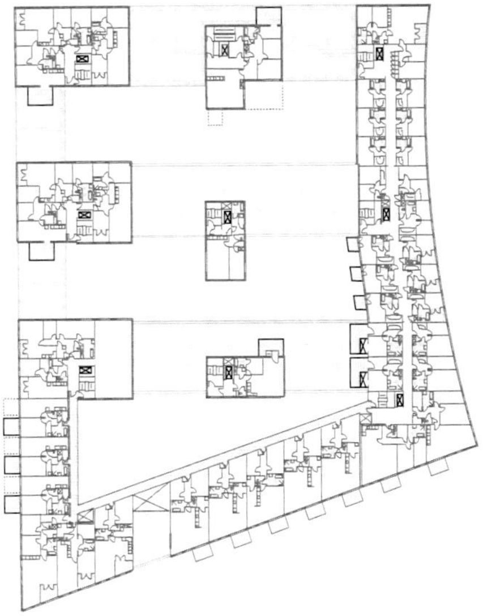

Level 3

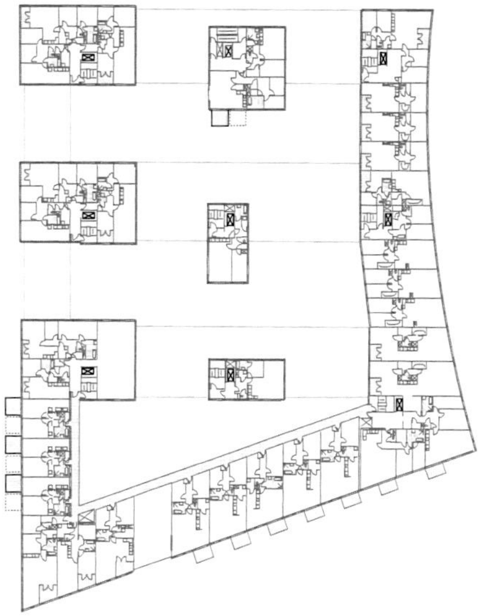

Level 4

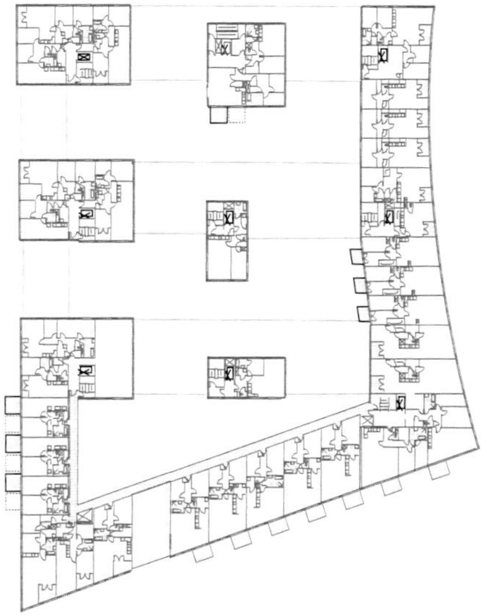

Level 5

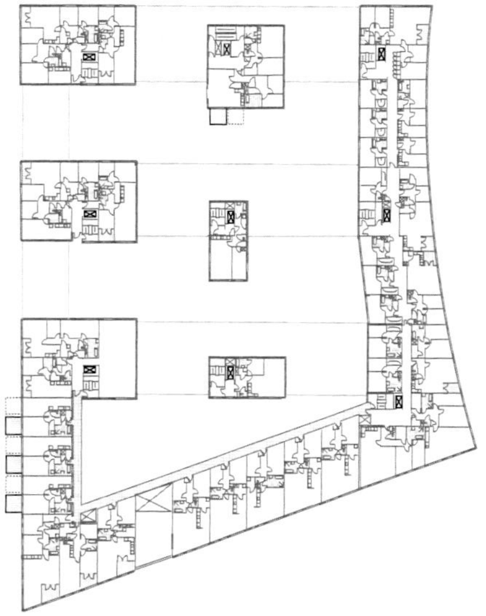

Level 6

0m 5m 10m 25m

1:1500

La Grande Cour floor plans: levels 7 to 10

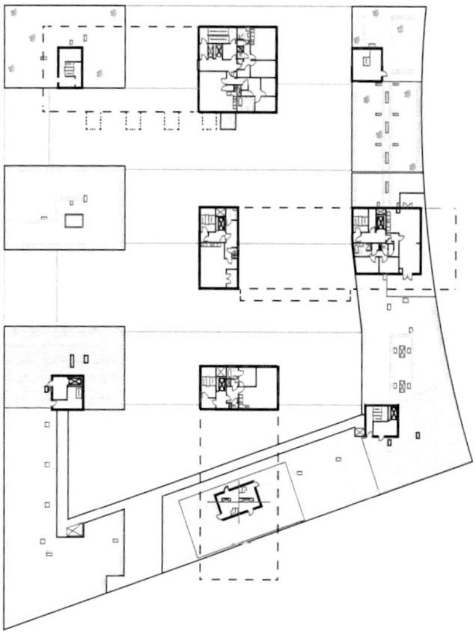

Level 7

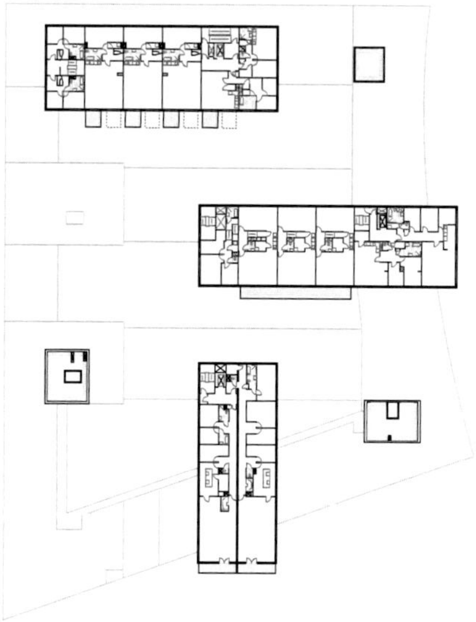

Level 8

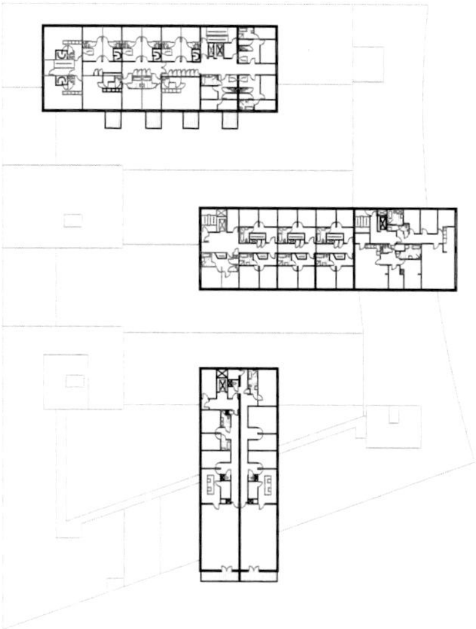

Level 9

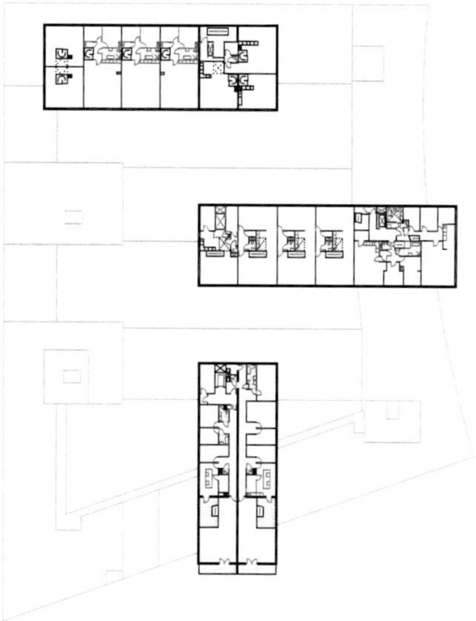

Level 10

0m 5m 10m 25m

1:1500

La Grande Cour floor plans: level 11 and roof

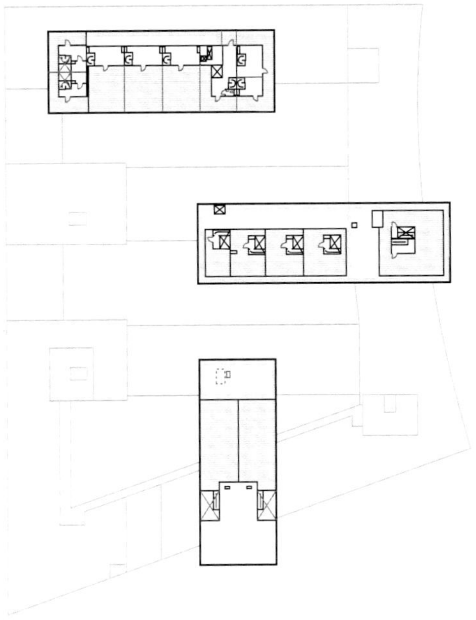
Level 11

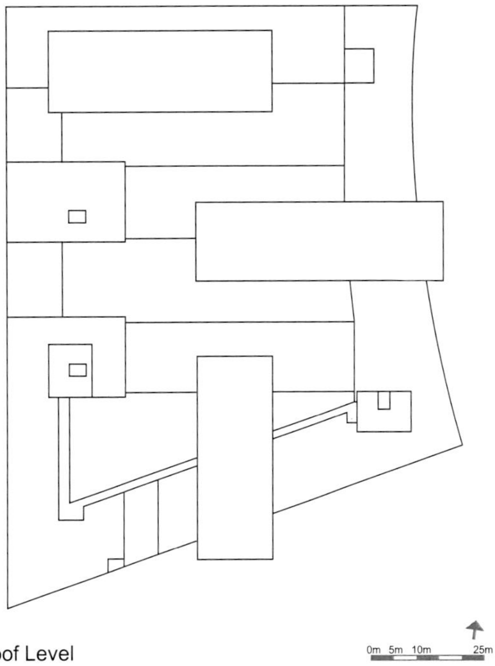

Roof Level

1:1500

The site itself slopes up at least a full level from the quay on the west to the main street to the east, Westerdoksdijk. To the north, fronting on the sloping, narrow Winthontstraat (Winthont Street) is the first of the adjacent buildings, and to the south is the Westerdoksplein or plaza. There are two levels of underground storage and parking for automobiles and bicycles, accessible from the nine elevators.

Organization, Appearance, Form, and Massing

The housing within the La Grande Cour complex is oriented to the three courtyards, each of which has a different character. Within the residential buildings are several community spaces, some with roof deck access. Ringing the building on all four sides, the commercial areas are accessible at the street level. The two-level garage beneath the building holds 263 cars.[10]

The complex is a perimeter block scheme, but with three courtyards. The long linear slab on the east comprises three interconnected, eight-story buildings, each entered via a different courtyard (see diagram page 167) with three skip-stop maisonette buildings in the middle sandwiched by vestibule flats on each end. Inside the block, rowhouses line the courtyards. Forming the west side and creating gateways into the courtyards are towers of vestibule flats that support suspended gallery-accessed flats. Gallery flats also frame the southwest corner and the south side. Finally, the three periscope buildings are inserted into the middle of the courtyards.

The Four Façades

Each façade is unique, reflecting its particular situation although all four facades have commercial space at the ground level. The deep red/purple brick west façade along the quay is not a solid flat plane, but where the middle and north courtyards are located has substantial openings that allow light in the courtyards. Additionally the volume of the L-shaped north periscope building on the north end is differentiated from the lower building mass by spatial separation with an open level below and by its blue-gray brick envelope. The balconies on the west side serve as counterpoint to the largest area of brick on the southwest corner.

The north façade is not very visible since it faces the very narrow Winthontstraat. Here the six- or seven-story dark brick cluster apartment buildings hold the corners, connected by the rowhouses along the bottom, also in brick. These are penetrated above the second story by the narrow blue-gray-paneled north periscope building that opens out at the top in the large mass of skip-stop, units, flats, and penthouses. The entrance to the garage is on this side.

Along the windy IJ River, the east façade is more austere than the other sides of the building, with its red-purple brick plane rising up six levels and the symmetrically placed white window and door surrounds defining the middle section. Here, the units' exterior spaces are loggias whose glass windows protect the outside areas from the wind and maintain the smooth building façade. However, the austere look is subverted by the middle periscope building that teasingly hovers over it.

West façade
© Scagliola/Brakkee

North façade

East façade

South façade

Section through the south courtyard showing the change in level (looking south)

The south façade faces the Westerdoksplein. It is enlivened not only by its two openings, but also by the glazing pattern of the commercial spaces at the base of the building, the overhanging south periscope building, and the balconies along its eastern end.

The Three Courtyards

Each courtyard accommodates entrances to rowhouses and to three multi-family buildings: a vestibule building to the west, a periscope building in the middle, a the linear building to the east as well as gallery flats along the east side, and in the south courtyard, the south side as well.

The South Courtyard

The southern courtyard, designed by MVSA, is elegant and cool. Shaped like a triangle and sloped, it is paved with white concrete and includes two gravel and shrub gardens, one with a tree. Unlike the other courtyards, this courtyard is entered exclusively from the quay to the west with a white concrete ramp up to the main level. At the main level the courtyard window opens to the south toward the plaza and canal that are one level lower.

The entry for the south vestibule building that serves both vestibule flats and gallery flats is at the level of the quay. At the main level are the entries for the south periscope building, the south linear building, a community space, and five rowhouses. The four of the five rowhouses on the north side of the courtyard are

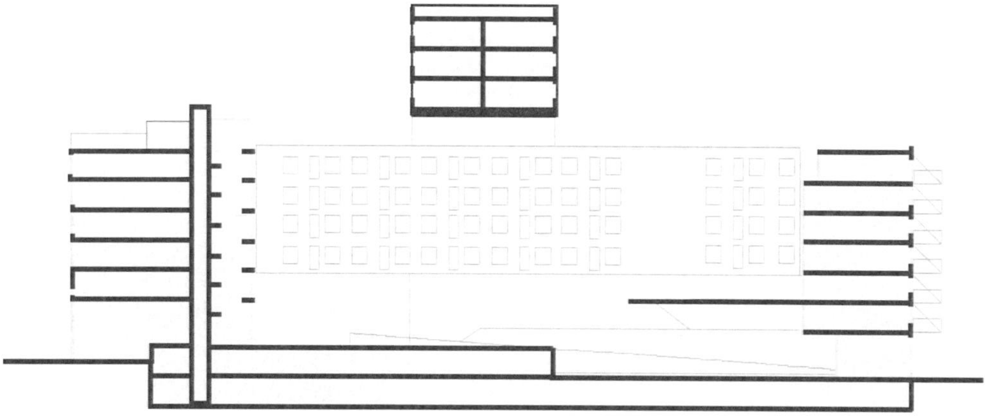

Section

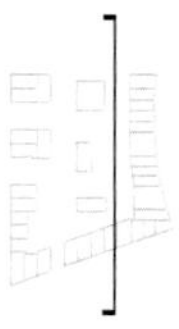

South courtyard showing reflected light
© Dirk Verwoerd

South courtyard with gallery flats on two sides
© Jeroen Musch

oriented exclusively to the south. Their entry areas are differentiated from the courtyard circulation path by a change of grade and a parapet wall. The fifth rowhouse, entering directly from the plaza on the south side, is oriented both north to the courtyard and south to the plaza.

At the east end of the courtyard, the elevator for the south end of the linear block connects to skip-stop maisonettes and cluster apartments that overlook the main street and IJ River to the east, and the courtyard to the west. The linear south block encompasses twenty dwellings.

The entrance to the south periscope apartment building is opposite the south opening and adjacent to the north rowhouses and the community space. The south periscope accommodates one apartment per floor between levels 2 and 7, and six penthouse units on levels 8–10, for a total of eleven units. The penthouse dwellings project over the south façade of the complex with orientation for the west units in two directions or those on the east in three directions.

The south vestibule building serves as entrance for by far the most units in the complex: seventy-seven dwellings and one community space on the third floor. Of these

Middle courtyard at La Grande Cour
© Scagliola/Brakkee

residences, sixty-one are gallery apartments along the west side of the building and sixteen are cluster or vestibule apartments. Most of the gallery apartments look west over the quay and canal, to the east toward the courtyard, with west-facing living rooms. The cluster apartments have windows in one or two directions, toward the west, north, and/or south.

The south vestibule building is skinned with a white concrete surface up to the second or third levels and above that a shiny gray metallic panel. The galleries are wood decking, and those along the south side of the courtyard have silver screens.

While the south periscope building has a dark gray metallic façade, the rowhouses and the north side of the skip-stop building are a very light gray concrete, with white concrete walkways. Due to the reflectance of the pale colors from the south opening, the courtyard appears light-filled (see photos at on page 175).

The Middle Courtyard

The middle courtyard, designed by de Architekten Cie, is notable for its orange ground surface and wooden decking. The gray wooden stairs come up from the quay and at the top transform to become a wooden walkway surrounded by the orange resilient rubber–covered ground plane that seem intended to delineate the outdoor area associated with the rowhouses. The linear east building is paneled in light blue, while the surface of the remaining buildings is reflective gray metallic panel. In all the buildings the units are expressed as a collection of windows in panel surrounded by white edging that provide a sense of human scale in what is otherwise a very cool courtyard, despite the hot orange carpet.

In addition to the three residential multi-family buildings that enter from the middle courtyard are eleven rowhouses that are oriented exclusively to the courtyard, facing north or south.

Balconies of skip-stop units in the middle linear block under the middle periscope building

North courtyard
© Jeroen Musch

On the east end, the middle linear building bridges the courtyard at the second courtyard entrance, and is entered beneath its overhang (see photo below, left). Between levels 2 and 6 it is connected via fire doors on each level to the adjacent northeast and southeast buildings to create one long, interconnected building that protects the three courtyards from wind blowing from the IJ River. The middle linear building consists largely of skip-stop maisonettes on levels 2–8 that enter from a double-loaded corridor. The only exceptions are two flats on level 6. Beginning on the eighth floor, this building is connected to the middle periscope building, providing emergency stair access for the skip-stop units that enter on level 9. Where it ties to the periscope structure, there are not only two flats per floor from the eighth to the tenth floors, but also one penthouse apartment on level 7 and two on level 10. In total, the middle linear building includes twenty-three units.

The middle periscope building that enters from the north side of the middle courtyard has a single one-bedroom flat on each level from 2 to 7. As just noted, starting at level 8 the design links to the middle linear building's emergency stair and includes seven skip-stop maisonettes for a total of thirteen units in the building. The flats have light on three sides, but exceptionally in this project, do not have outdoor space. At the ninth floor a corridor serves seven three-bedroom skip-stop maisonette units, four of which have access to roof gardens at the eleventh floor, and three of which have decks at the eighth floor. Also at the ninth floor, the periscope building connects to the middle linear building with access to the two apartments per level, as mentioned earlier.

The middle vestibule building on the northeast corner is accessible from the middle courtyard at the level of the quay of the courtyard. On each level from 1 to 6 are three cluster flats, and on levels 1 and 2 there is also one gallery unit, for a total of twenty units in the building. As with the north vestibule building, at the west end of the middle courtyard, the two gallery apartments, one atop the other, bridge and enclose the courtyard on levels 1 and 2. These gallery units are located between the middle and north vestibule towers, and have access from the middle vestibule building.

The North Courtyard

The rectangular north courtyard by architects Heren 5 Architecten is designed as a series of paved terraces from the quay at the west end up to the street level at the east side, with entrances both from the Westerdok quay and from the main street, Westerdoksdijk, like the middle courtyard. The gray stone paving is complemented by the blue-gray color of the surrounding buildings and the white of the benches and planters, creating a sense of peace. Although the blue-gray brick north side of the court is ten stories tall in the middle and at the west end, surprisingly, it is not oppressive, but delightful to be in the courtyard, because the wall has openings that allow light to enter the courtyard, and the periscope overhang does not lie directly on the buildings below, but floats over them.

As with the other courtyards, the social rowhousing is entered from the courtyard, but in this instance, there is a raised 4-meter deep wooden deck along the fronts of the houses on both sides. The deck, wide enough to inhabit, is at the same level for the entire courtyard, and separated from the terraces by a planter or bench. As the terraces descend, the planters and benches remain at a height consistent with the deck, and the level difference with the terraces is mediated by stairs. This layer of territorial differentiation between the public walkway and the private deck supports residents' control of their exterior space, as demonstrated by the plants that several residents had placed by their front doors.

Similar to the middle linear building, the north linear building faces the main street, Westerdoksdijk, with its vertical circulation core entered underneath its building mass directly from the court. At the east end, the north linear building incorporates skip-stop maisonettes with double-loaded corridor access on the third and sixth floors, but it also has a cluster flat at the north end on every level from 2 to 6. The connection to the other linear buildings was described earlier.

The north periscope building, entered from the main level of the north courtyard, consists of pairs of east and west-facing two-bedroom apartments on levels 2–8 (with the exception of a penthouse unit with roof garden on level 4), and four levels of housing at the top, mostly reached by a double-loaded corridor on level 9. A level-9 link between the periscope tower and the north vestibule building provides emergency access to the eleven skip-stop maisonettes on levels 8–11 and the two flats on level 8. The north periscope building contains twenty-four units, of which twelve are flats, three are maisonettes, and nine are maisonette penthouses.

Virtually identical to the middle vestibule building except for its internal connection to the north periscope building, the north vestibule building at the northeast corner of the north courtyard frames the north courtyard with two gallery apartments that bridge and enclose it on levels 1 and 2, between the vestibule towers. Again, three cluster flats are on each level from 1 to 6. This tower marks the northeast corner of the entire complex. As with the east central building, the flats and the gallery units are accessible from the lowest level of the courtyard, at the level of the quay, for a total of twenty units in the building. The connection to the north periscope building is through the extension of the elevator core to the double-loaded corridor on level 9.

La Grande Cour unit plans: A

Housing and Unit Typology

La Grande Cour has five basic housing types: (1) rowhouses, (2) back-to-back rowhouses, (3) skip-stop maisonettes, (4) cluster apartments, and (5) penthouses. The number of apartment variants is about twenty-two, developed based upon access and orientation.

Rowhouse

All thirty-six of the rowhouses were designed as social housing units accessed at ground level, and opening onto courtyards. Their exterior space is found directly outside the entrance space delineated differently in each courtyard. In the north courtyard are five large three- or four-bedroom rowhouses that have both north and south orientations. There is one three-bedroom dwelling with north and south orientations in the south courtyard. The remaining rowhouses are two-bedroom back-to back dwellings oriented either north or south.

The six single-orientation dwellings in the north courtyard face north with the wide decks as outdoor space. In the middle courtyard, five face south and six face north. Although these units appear to have clearly delineated outdoor space in plan with a change in material, because there is no clear boundary in the third dimension, the territory is not defensible. The south-facing rowhouses in the south courtyard have a separate entry territory but without sufficient space to accommodate a plant or a seat. Compared to the other units in the complex, those rowhouses that have access only to north or south are deficient, as are those without a clear outside space. Certainly, assigning a courtyard to each architect made it difficult to create a design that would have given these units two orientations. However with cooperation, it would have been possible to create a cross-over design to allow dwellings to have two orientations with the entry and living level facing one courtyard and the bedroom level facing another.

Middle periscope building with vestibule flats in vertical tower and skip-stop units suspended above rowhouses at the base © Jeroen Musch

Skip-Stop Maisonette

Skip-stop maisonettes are found in two locations, the east linear building and the north and middle periscope buildings (see plan b, page 179). The skip-stop maisonettes along the east side of the linear block, most having two bedrooms, are organized within the three connecting eastern buildings. Varying in depth due to the curve of the main street, they face the river on the east and one of the courtyards to the west. Most have living rooms toward the east with loggias as outdoor space, responding to the wind from the river. A few of these units have balconies facing south toward the courtyard, and a very small number have south-facing loggias. The maisonettes are all entered from double-loaded corridors at levels 3 and 6 in their respective buildings. As explained earlier, for fire access purposes, the corridors along the east side are connected to each other. Different paint and carpet color in each building supports orientation.

The three-bedroom maisonettes at the tops of the north and middle periscope buildings are oriented north-south and have south-facing balconies (see photo above right for the north periscope building). Nine of these twelve maisonette dwellings are penthouse units with rooftop terraces.

Vestibule Flat

Sixteen two-bedroom vestibule or cluster apartments are located at the north and south linear buildings on floors two through seven (see plan c, page 179). Being corner units they have two orientations, toward the building exterior rather than toward the courtyards, with loggias for exterior space.

A second set of cluster apartments, also two-bedroom units, is located along the west side of the complex in the three vestibule towers that serve as gates to the courtyards. There are three per cluster for a total of fourteen, because one is omitted on the sixth floor of the south vestibule building to create a community space. Two in each group orient either north and west or south and west. The third has three orientations: north, east and south. Their outdoor space is a loggia.

The final types of cluster flats are in the periscope buildings. These vary greatly, having from one to three bedrooms and being located in one or two corners, thus with orientation in at least two directions, and sometimes three or four. Those units in the trunk of the periscopes (levels 2–7) are unique to each building, with one-and three-bedroom dwellings in the north building, and one-bedroom dwellings exclusively in the middle and south buildings. While the one-bedroom units in the south building have south-facing loggias, those in the north and middle buildings do not have exterior space. The three-bedroom units in the north periscope building have south-facing balconies.

The vestibule flats located at the top of the north and middle periscopes on the east side are large dwellings with a river orientation, but without exterior space.

Gallery Flat

The gallery flats are located along the south and west corner of the complex with access from the south vestibule building and the south linear building. Galleries on levels 3–6 are connected both to the south linear building and to the south vestibule building, so that the flats on these three levels can enter from either building.

Flats that are reached through the south vestibule building form the southwest corner of the building, with the lowest level bridging over the south vestibule entrance. The first and second galleries are entered exclusively from the south vestibule building, and those above are on the connected gallery. Facing west on the southwest corner are eighteen two-bedroom flats three to a level on floors 1–6. The living rooms of the three south-facing units look toward the Westerdoksplein with bedrooms also facing south, or north into the gallery and courtyard.

On the southwest corner itself, for all six levels are six two-bedroom corner dwellings with bedrooms and living

La Grande Cour unit plans: B

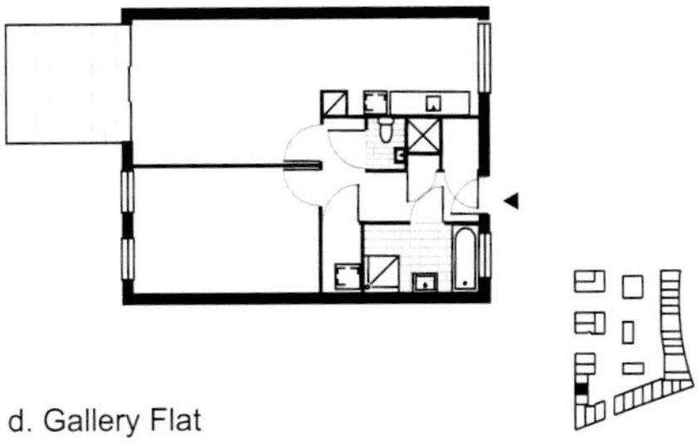

d. Gallery Flat

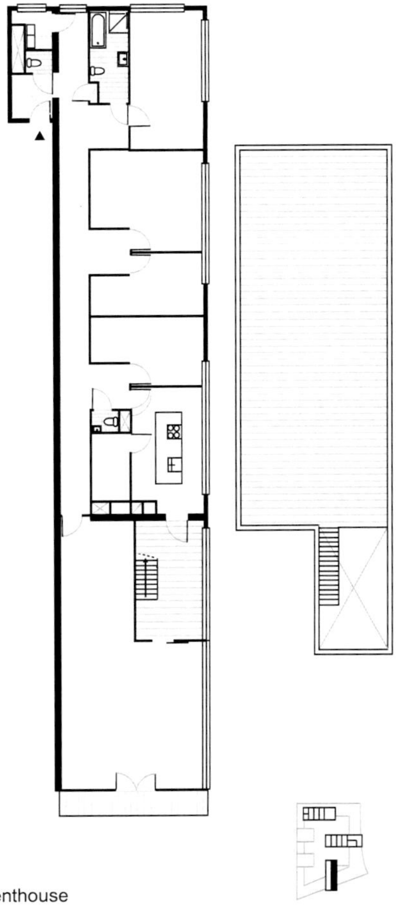

e. Penthouse

1:250

Penthouse living room
© Jeroen Musch

room facing west towards the quay, and a few extra living room windows looking south as well.

Again entering from the south vestibule building, on floors one to six, along the south next to the corner unit are another three gallery flats on each level for a total of eighteen. The living rooms of these flats overlook the Westerdoksplein to the South. Two are two-bedroom units, and between them lies a one-bedroom dwelling. Additionally, on level 1 there is a nineteenth one-bedroom gallery flat next to a rowhouse. All nineteen have access to loggias for exterior space.

The final set of fifteen gallery apartments, primarily accessed from the south linear building, are on levels 3–6, five apartments per level. These are all two-bedroom flats with living rooms and balconies that look south over the Westerdoksplein, and bedrooms that look north onto the gallery and also south toward the courtyard.

Penthouse Flat

The penthouse is the unit type with the most ambiguous unit definition because it is not defined by its room organization but designates units that have roof terraces, whether decks on someone else's roof or a garden on one's own roof (see plan e, page 181). In this complex, because of the large amount of roof deck space, even some very small units are penthouses.

Having discussed penthouse maisonettes in the section on that building type, here we focus on the nine dwellings in La Grande Cour that have their interior space on one level as flats, although their roof gardens may be on the same level, a level above, or a level below their inside space.

The four most typical penthouses are found on the eighth and tenth levels of the south periscope building. These upper income dwellings are positioned two to each floor: two 4-bedroom units on the east side facing the river, oriented north, south, and east; and two 3-bedroom dwellings on the west side facing the city, oriented south and west. While all of the dwellings have large south-facing balconies, those on the eighth floor have roof gardens one level down, on the roof of the gallery flats, and those on the ninth floor have extra decks on the ninth floor and roof gardens one level up on their own roof.

La Grande Cour space syntax gamma diagram

In the middle periscope building, a pair of two-bedroom penthouse flats on level 10 have roof gardens above on level 11.

A final set of three penthouses has roof gardens at the same level as the interior spaces. On level 3 there is a two-bedroom apartment in the north periscope building and a one-bedroom dwelling in the south periscope buildings; both have living rooms and roof gardens facing west over the social housing. And on level 7 there is a flat in the north central building that has a roof garden on both sides, over the northeast and north-central maisonettes.

Syntactical Structure

The gamma diagram for La Grande Cour is unique among the diagrams partly because of its large number of threads or circulation paths, and also because of the small number of units in large fan-shaped circulation patterns compared to other diagrams, in part due to the periscope buildings that have only a few units on most levels, but also because of the breakdown of the corridors into segments for exiting. It is especially interesting to compare the diagram with that of Silodam, the other highly dense complex with more than 300 units per hectare, which has three threads in contrast to the nine threads in La Grande Cour. There is similarity in the high level of interconnectivity, but the average housing thread for La Grande Cour has twenty-eight units in comparison to forty for Silodam. The threads (representing units using one elevator) in La Grande Cour have between fourteen and fifty-two units, with the three periscope threads having eleven to fifteen units on a thread, and one large thread that has fifty-two units. Thus the dwellings at La Grande Cour are more widely distributed than those at Silodam and most of the other projects.

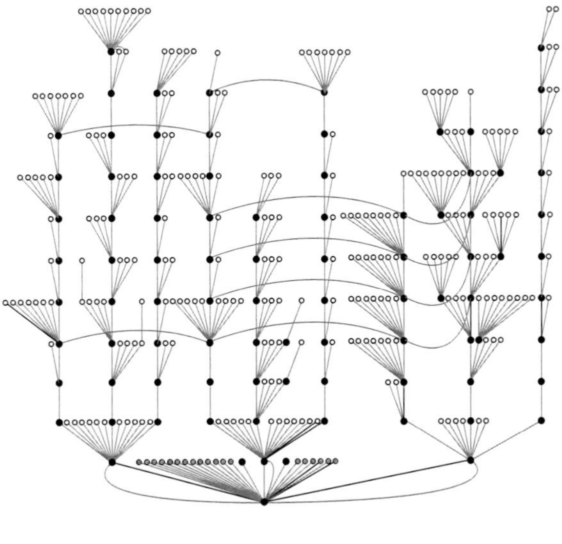

The most distributed building in the complex is the middle periscope building, with seven units in the tower, one unit per floor, and six luxury units on the top three levels, each accessed from a vestibule shared with one other unit (see 4th thread from the right). This building has one elevator for thirteen units. In contrast, the most collected building is the south vestibule building with gallery and vestibule flats (see 2nd thread from the right). This building has eleven collections of units (accessible from one elevator stop) with an average of five units per collection. These represent seven to eleven vestibule and gallery-accessed flats per floor, on each of five floors, for a total of fifty-two units using that elevator.

The varying character of the threads in the space syntax diagram reveals the complexity of La Grande Cour. The three periscope tower threads create a unique syntax pattern with few units in the towers, and in two cases, a large number of units in a fan at the top. The threads of the buildings that form the eastern wall show the alternating pattern of the skip-stop, with large fans on some floors and a few entries on the others. The buildings along the western side of the project have threads with a repeated pattern of gallery-accessed fans of a similar size. Despite the complexity, however, the organization of the building is clear, with a number of dispersed units as well as units collected in fans in many threads and on many levels. The circulation is very distributed, with a total of nine elevators for average of about twenty-eight units per elevator. This average is very close to the distribution of elevators in the less dense buildings and contrasts greatly with Silodam, whose ratio is fifty-one dwellings per elevator.

Social Principles

As the developer Rein Willems said, this building, like the others in this book, is a social experiment. There is a combination of rental units and owned units, and while there is a mix of social housing, housing for people of moderate means, and luxury housing, it isn't clear from looking at the buildings or the dwellings from the exterior which units are for which groups. The rowhouses, designed for social housing, are uniformly entered from the courtyards, thus largely independent from other units, connecting only in the courts. There is some integration of units, with several buildings mixing either social housing and moderate-income units and others mixing moderate and luxury units. But there is no building that combines social housing and luxury units. Thus the strategy for income mix is to combine segregation by income for some dwellings (rowhousing) with integration between two adjacent income groups, either high and middle (periscope buildings and the eastern buildings) or middle and low (gallery-accessed buildings).

When I visited the site in March of 2010, and more recently in 2017, there was a noticeable difference in the design and signs of personalization of the exterior areas associated with the rowhouses in the three different courtyards. In the middle courtyard by Cie, where the public walkway and the good-sized areas adjacent to the rowhouses are only differentiated by a change in material, there was no personalization, although to some extent this may be due to this courtyard being inhabited by temporary international Deltalloyd visitors.[11] In the south courtyard by MVSA, where the territory outside the units is limited to an access walkway similar in material to the public walk, again there were no signs of personalization. However, in the west courtyard, designed by Heren 5 Architecten, planters are located to create a defined territory of about 4 meters between the walkway and the houses, and there is a change in level and of material. Here, while personalization was limited, people have added plants in containers and planter boxes to further differentiate their terraces. In this third courtyard, the design created territories that can be inhabited.

Lessons and Conclusions

Perhaps the most important lesson from La Grande Cour is that of the cooperation between planners and architects when the error in urban design calculations was discovered. The city's criterion for a limited building height had to be addressed at the same time that the density of the project had to be maintained. By working together, they reached a solution that created an iconic design.

Another lesson is the ability to experiment with a population mix that was enhanced by having a special site. On a highly desirable site in Amsterdam, with a downtown location in the West Docklands and views on all sides, it was possible to mix income groups in a single project. The developer undertook La Grande Cour, a highly visible project, with full awareness that this would become a national landmark. Their willingness to invest in the project has created a successful experiment.

The project has been carefully managed, perhaps because its visibility affords a branding opportunity. Nevertheless, it was built during the recession and certain concessions had to be made, namely selling the social housing to raise the capital to build it. But it is remarkable that such a site should be available to all income levels, and a testament to the city planning at the turn of the twenty-first century that dictated both the density of the site and the mix of incomes served.

La Grande Cour from the southeast

The appearance of the project and its complexity, partially due to having to compensate for the error in estimation made at the very beginning, contribute to its unique status. The design supports the informal surveillance, socialization, and strong identity that build community within the project. Generous use of elevators reduces the numbers of people using each one. Creating small groups of units within the project, and access to one of the community spaces supports neighbors knowing each other. Including commercial spaces enhances the street life. Furthermore the iconic design with the periscopes and the division of the project into three courtyards serves to create a strong urban identity for the project.

Notes

1. Ton Schaap (City of Amsterdam, Urban Designer), Interview, Amsterdam, March 23, 2010.
2. Rein Willems (Bouwfonds, Regional Director, developer, La Grande Cour), Interview, Amsterdam, March 25, 2010, 1.
3. Rein Willems, Email Communication, September 24, 2016.
4. Rein Willems, Interview, 2010, 2.
5. Ibid.
6. Jeroen van Schooten (MVSA, architect, La Grande Cour), Interview, Amsterdam, March 5, 2010, 1.
7. Rein Willems, Interview, 2010, 3.
8. Roger Sherwood, "Westerdok Island/La Grande Cour," Housing Prototypes Website, 2002, housingprototypes.org /project?File_No=NL 017.
9. Rudy Uytenhaak and Jeroen Mensink, "Westerdoksdijk Amsterdam," *Archimed*, April, 2006, www.archined.nl/en/reviews/engels/westerdoksdijk-amsterdam/.
10. MVSA Architects, Website, 2016, www.mvsa-architects.com/project/detail/29/la-grande-cour.
11. Jeroen van Schooten, Site visit, Amsterdam, May 25, 2017.

9 Silodam

352

Silodam south façade and the quay

Location of Silodam in the Netherlands

Location of Silodam in Amsterdam

Location of Silodam in the West Docklands

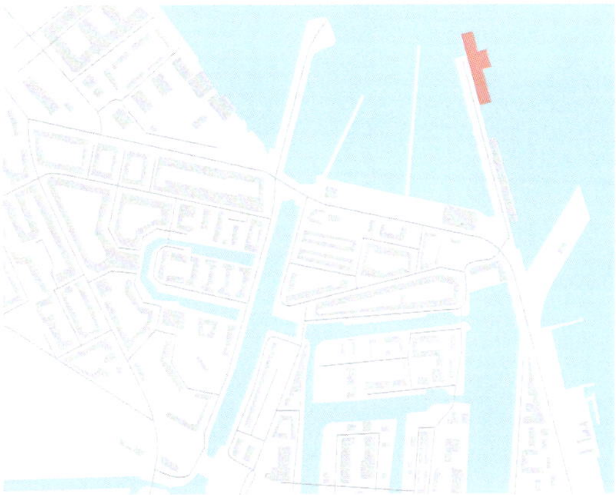

Completed in 2003 and designed to look like a container ship ("a modern warehouse to store people"), Silodam was the first new housing project built in the West Docklands area of Amsterdam, an area near the train station that historically was a warehouse and shipping area.[1],[2] As a part of the changes to the city since the development of container ships, the Westerdok has been transformed into a major housing district that includes the nearby La Grande Cour project presented earlier in this text. The construction of Silodam initiated the process.

Silodam was selected for this book because of its unusual design and its high density.[3] Although the subsequent projects built in this part of the city were required to meet a 200–300 unit per hectare density, Silodam's density is double that, at 600 dwelling units per hectare (245 units per acre).[4,5] The architectural firm MVRDV, known for its experimental approach to design, approached this building as a laboratory for the development of different housing and unit types. As stated by Nathalie de Vries, co-founder of MVRDV,

> MVRDV brought different house types together within the Silodam block: families, older people, people with many different hobbies, attitudes and lifestyles. And they're all united in one building. And that's also why we keep the exterior of the building simple. It's a big block lying in the water.[6]

The project served a variety of clients. As identified by the project architect Frans de Witte, these were the commercial developer (Rabo Vastgoed), the social housing developer (De Principaal—now De Key), and a group of senior residents that wanted to live together "and have twelve houses. . . . They were involved from the beginning, and we spoke to right them from the start."[7]

Silodam as container ship

Site, Context, and Building Functions

The wharf site for this project included two historical grain elevators, built around 1918, that had been abandoned for about thirty years, with the exception of some squatters who occupied part of the buildings. The squatters were originally included in the project discussions, but declined to participate when some of their requirements could not be accommodated. The building itself is in the water, and the associated parking is under the wharf.

Building in a water site created unusual conditions. The structure not only had to resist high winds and water, but also had to be protected from boats navigating the river that might go off-course. That area of the river is a primary transport zone for large ships. To prevent the building from being destroyed by wayward ships, a protective, shallow, underwater barrier was built in the river. Also the building is designed to withstand pressure

Silodam main entrance

Lobby and stairs to public deck

from ice that the city found tends to occur at the site. "The whole first row of columns can actually collapse [if necessary] and the building will stay upright."[8] Wind pressure and salt corrosion were also important considerations.

> Silodam is on a really windy spot—very close to the sea. We had to consider wind pressure and salt corrosion. We limited the amount of openings and the width-depth ratio was limited, to make sure we wouldn't have wind or water in the apartments.[9]

Furthermore, the soils did not allow a deep enough parking structure under the public wharf to accommodate the need for parking, so the decision was made to create an automated parking structure that would allow more cars in a much smaller space. Working with the city, the project was also able to include a small marina underneath the building.

As mentioned earlier, the river site generated an aesthetic response as well. The architects saw the building as a kind of container ship/warehouse that would house a maximum number of different kinds of units. Thus they designed it to fit the maximum building envelope. The housing responded to a 30/70 mix of lower and higher incomes, with a great mix of unit designs that "provided a range of living types," seventeen in all.[10] Concern that there might be difficulty selling all of the different types of housing proved unwarranted, as the site location near the train station that provided every unit views of the city on one side and views of the river on the other provided far greater interest than could be accommodated. "In the end we could have sold anything; 3,500 people wanted a house."[11]

The city of Amsterdam wanted working space in all of the buildings on the site, as well as public space associated with the housing.[12] Silodam was therefore designed to accommodate a restaurant and office space, with even more workspace in the two adjacent buildings.[13] Storage is located on the lowest level of the main building block, and bicycle parking is next to the building entry on the wharf. To address the required provision of public space, the project includes a large open deck overlooking the river located on top of the protruding restaurant on the east side of the building, accessible to the public by passing underneath the building and climbing a ramp or stairs.

Public deck from above

Silodam circulation diagram

Organization, Appearance, Form, and Massing

The building does indeed resemble a container ship with the units stacked up in four sections, each type of unit with its own architectural expression, and all units connected to the central circulation system accessible through three elevator cores. "We have corridor houses, balcony houses and houses with direct access from the dam, because we wanted diversity in the block."[14]

The four stacks of units are visible along the length of the building, and the different unit types are uniquely colored and fenestrated so that they will be recognizable. "We thought it was really important that you could say 'I live there—in the orange houses.' "[15]

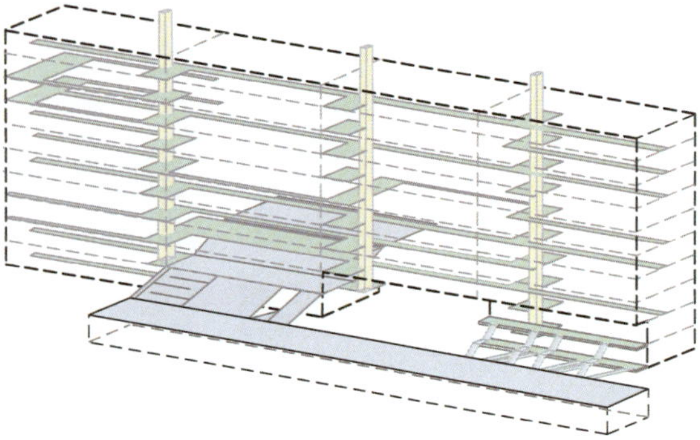

To make the great variety of units work within the building, the designers carefully located the plumbing and ventilation shafts.

> We tried to make the units as flexible as possible. Shafts, toilets, bathrooms are all positioned at a certain place. Where we couldn't fit the ideal arrangement [for a unit type] we put them somewhere else.[16]

In many cases there was opportunity for purchasers to arrange the dwelling spaces themselves. "Some people put the kitchen there, some people the living room, there."[17]

The circulation system, or routing as the architect calls it, was a central concept in organizing the building. The building is arranged as a set of neighborhoods, all connected into one large community with a single point of access to the three elevators. Each type of housing forms a neighborhood recognizable on the exterior by its own façade design. Furthermore, the clusters of units have various corridor or gallery-accessed design. All residents have access to the entire building, and to a number of shared spaces. Some of these were planned by the architect, such as the level 8 north deck observation area; other spaces, extra leftover areas in the building, were adopted by the residents themselves for such purposes as a book library, toy library, workout area, and meeting space after the building was occupied.

> We also wanted to create a routing inside the building that everybody would use with a viewing point on the 8th level. If you have people over [to visit] you go through the building to this public point.[18]

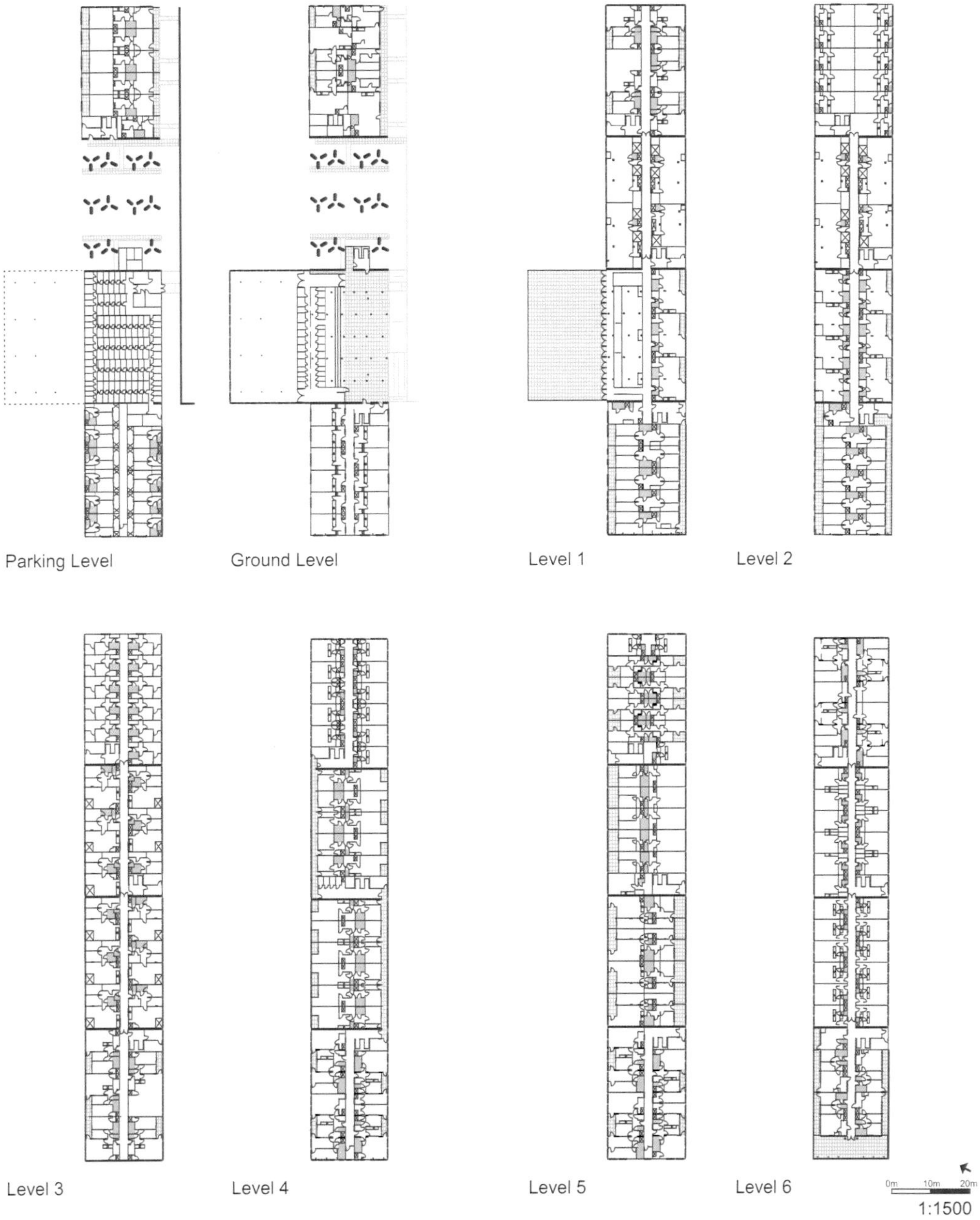

Silodam floor plans: parking to level 6

Silodam floor plans: level 7 to roof

Top of left column "Corridor"

Top of right column "Gallery"

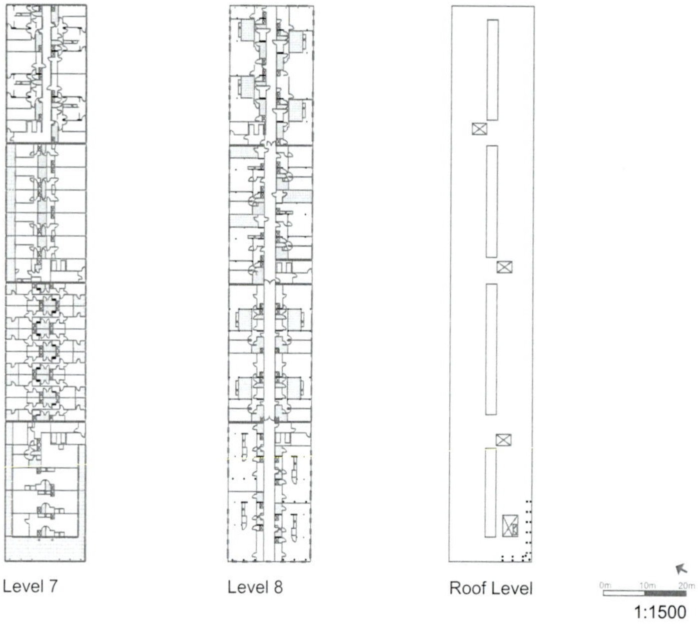

Top caption left column "Community observation deck on level 7"

Bottom caption left column "Toy library"

Bottom capation right column "Workout room"

Silodam is designed to be seen from a distance, as the primary views are from the water and from streets located at a good distance from the building. As a building that is ten stories high with a length of about 1,300 meters, the building appears from the side as a massive block made up of many parts. From the ends, however, with a 20-meter dimension, its slenderness is evident. The large scale is moderated by being broken down into four parts, and by each layer of housing type having a different expression.

Housing and Unit Typology

Of the seventeen unit types in the building, three are presented here. These examples only hint at the experimentation with unit design embodied in this building. Such variety is achieved by five factors: the types of access (double-loaded corridor and single-loaded corridor/gallery access), the number of levels in a unit (one or two), the type of exterior space provided (loggia, balcony, exterior gallery, and atria on the top level), the width and depth dimensions of the units, and the arrangement of the interior spaces. According to Eric Amory, the developer: "In the beginning there were many more housing types. In order to get the building built and houses sold we cut it down to a smaller number."[19]

The design incorporates three types of outdoor space.

> Most apartments have loggias. Some have the gallery that is used as a balcony. On the top floor [there is the] patio house. And one type that has a sunken roof terrace in the A block. . . . Also [there are] houses with a garden for a corridor on the façade—almost 9 meters deep.[20]

Ventilation was an important consideration. All of the units have operable windows and a ventilation system that removes air from the bathrooms and kitchens. As described earlier, it was also important to address the windy site with well-built windows that would resist wind pressure, salt, and corrosion.[21]

The dwelling designs discussed next are a flat on level 5 that is part of the cooperative project; a skip-stop, cross-over maisonette that enters from a double-loaded corridor on level 7; and an atrium flat located on the top floor of the building.

Silodam unit plans: A

Gallery-Accessed Flat

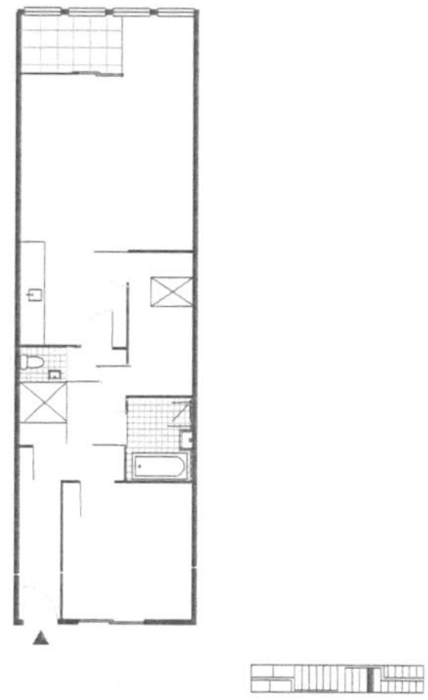

a. Gallery Flat

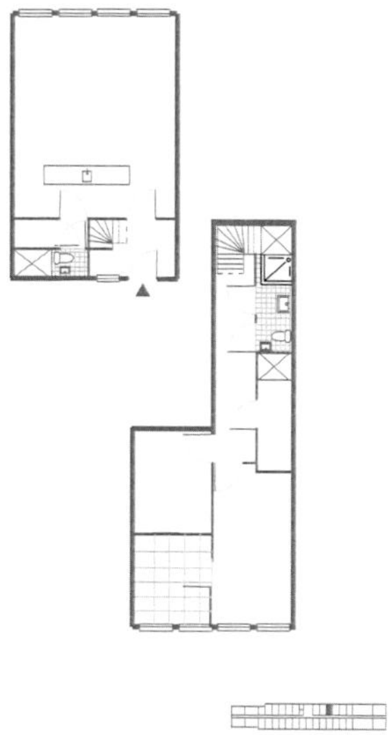

b. Crossover Maisonette

1:250

Cooperative housing for elderly: extra-wide glassed-in gallery space, gallery-accessed flat living room, and gallery stairs to units on upper level of the coop

Gallery-Accessed Flat

The flat shown here (see photos this page and plan a, page 196) is part of the cooperative area of senior residents who wanted to live together. This is the smaller of the two-unit designs for this group, very basic, designed to have an enclosed bedroom in the front or east side, a living room on the west with a loggia overlooking the river, and kitchen and bath on either side of a connecting corridor. The first residents were given the opportunity to tailor their units, so the illustration of the unit does not match the original plan that is shown. The resident of the apartment

photographed preferred an open plan design without a loggia. So in his apartment, while the kitchen and bathroom area are enclosed, the bedroom is open and connected to the entry and living room. Instead of a loggia and short wall to the west, the flat has a large wall with a picture window.

This group of cooperative flats has an unusual arrangement, with all residents of the collected units entering the extra-wide gallery on the fifth floor. The smaller lower flats are entered directly at the fifth-floor level, while the upper flats are reached by the gallery staircase.

Cross-Over Maisonette

Twenty-one skip-stop, cross-over maisonettes, with front doors on levels 5 and 7, have access to both the east and west sides of the building. The vestibule, WC, living area, and kitchen are on the entry level, with stairs rising to two bedrooms, bathroom, and loggia on the floor above. Eleven of these units have the living room facing west and bedrooms toward the east (see plan b, page 196); the twelve remaining face the opposite direction. All the dwellings on the fifth level connect to each other through a circuitous path. At the south end an elevator leads to the double-loaded corridor segment for these skip-stop units. In contrast, the twelve dwellings on level 7 enter from one segment of a long and straight corridor that runs the length of the building.

Atrium Flat

The entire the top level or ninth floor consists of sixteen atrium or deck flats. The corridor, extending the length of the building, is skylit and broken up into four sections, each of which comprises four units. Except for the four corner dwellings, the atrium units have exterior windows oriented only in one direction, either east or west. However, in the center or at the side of each is either an atrium that opens either onto the living room, or to a stair to a deck on the roof level.

Silodam unit plans: B

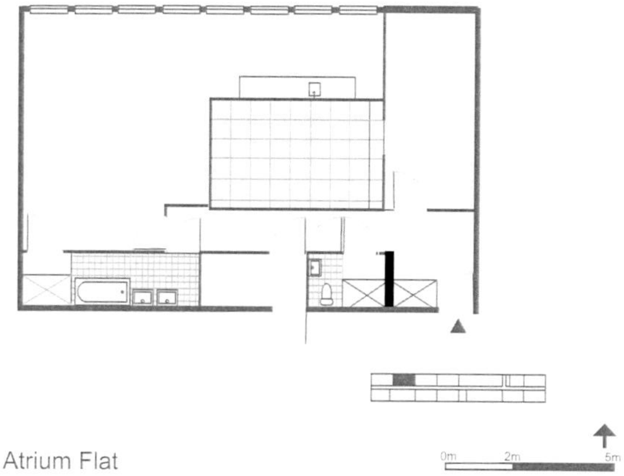

c. Atrium Flat

1:250

The flat represented here is in the southernmost group, facing east. One enters at the middle of the unit facing the atrium with the living room to the left and the bedroom and bathroom to the right. The central atrium opens onto the living room and bedroom. All three rooms—kitchen, living room, and bedroom—have large windows overlooking the eastern city view.

Syntactical Structure

As we have discussed, Silodam is organized like four layer cakes set side by side, each layer representing a different type of housing, and each layer having its own external expression. The four cakes are linked on several levels with corridors that in three cases go straight through the middle of the building as double-loaded halls from north to south, (levels 4, 7 and 9), and in one case (level 5) meanders from one side of the building to the other, as single-loaded corridors on east and west are linked through the elevator cores. Because a number of the dwelling types are two stories, the remaining five floors have hallways in only one or two parts of the building. The three elevator cores serve units on either side, but depending upon where fire separation doors are placed, they encourage use more by one side or the other.

Silodam space syntax gamma diagram

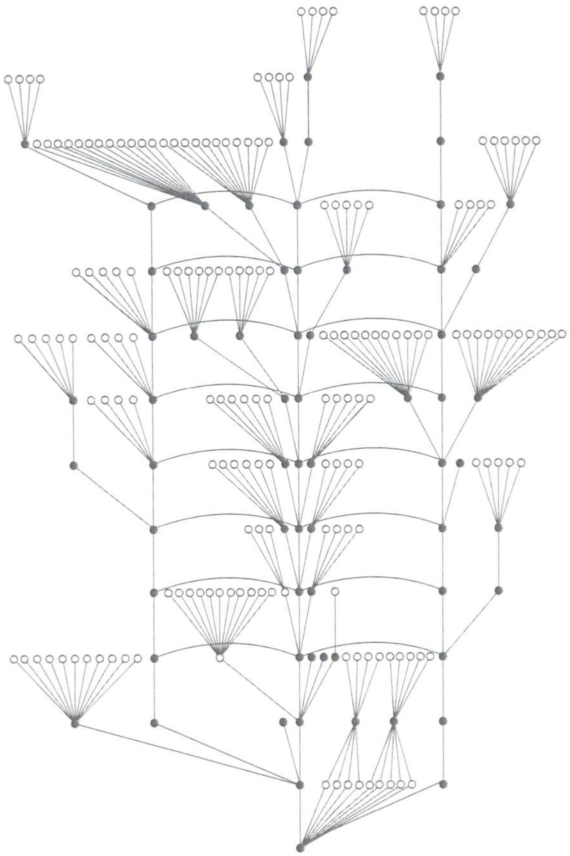

On the wharf: bicycle parking, entrance to social housing dwellings, and a garage entrance

The space syntax diagram shows that Silodam is one of the most highly internally connected of the eight case studies, and one of the least connected externally. Vrijburcht, for example, is similarly highly internally connected, but it also has strong external connections, with several entrances to its courtyard as well as unit entrances from the street. Silodam, on the other hand, has a single entrance for almost the entire building with ten dwellings that enter from the quay, and the businesses that enter through the lobby without connecting to the elevator system.

Another notable aspect of the syntax diagram is that in the north core (left thread), most units are directly connected to the elevator, whereas along the second and third elevator cores there are several long threads linked to each. This represents the degree of enclosure of the elevator core, which in the case of the north core opens within the corridor fire link. In other instances, the elevator core is enclosed and there are one or two corridors with fire doors between the elevator and the units.

The final intriguing threads are those at the far right side of the diagram that represent the social housing that enters directly from the wharf. Half of the units use individual ramps up over the water to connect to a shared deck a little below the level of the quay, and the other half employ stairs to go up to a shared deck a level above.

In sum, the syntactical structure of the diagram shows a structure that is deep, varied, highly organized around three elevator cores, and highly connected internally, while relatively closed to the exterior.

Social Principles

The project was designed to be a place where a variety of people would mix and includes social, middle-income, and luxury housing. According to Nathalie de Vries, there were various financing models for the different units.[22] Also, the design for a variety of incomes and unit types contributes to the diversity of the inhabitants. But integrating social housing with the other housing is controversial in the Netherlands. In this project, social housing constitutes 15% of the units. The one potential limitation to the diversity is that while there remain in the collective group a small number of social housing units within the building circulation, much of the social housing is entered directly from the wharf, does not appear to participate in the shared circulation spaces, and therefore may not be not a part of the internal community. Although it can be advantageous for families to have direct access to the out-of-doors, as with some of the other projects such as Carnisselande and La Grande

Silodam west façade

Cour, it would be unfortunate if the trade-off were to be outside the Silodam community.

In any case, the internal organization appears to be successful. Architect Frans de Witte notes that Silodam was designed to be a social place with "one entrance, everything connected to each other. It's one community."

> There were some leftover spaces, and because of the routing they use it. There was a leftover space they turned into a library, another became a small meeting space. These are places for everybody to use. The client actually saw the [great] potential of it . . . [and] afterwards . . . donated the binoculars [for the viewing point on the 8th level].[23]

When I visited the project in 2012, nine years after it was built, these narrow interstitial spaces were still actively used for a great variety of purposes, all pointing to the success of the architects' approach to creating community within the structure.

Lessons and Conclusions

One lesson to be learned from this experimental project is the degree to which site choice can be a determining factor for success. When a site is ideal, near the city

Silodam in its context, seen from the east side of the river

center, on the river with views in every direction, and made available at a low cost, even when the neighborhood is not yet developed, an experiment is more likely to succeed.

At the time of construction, however, success was not ensured. The developer Eric Amory pointed out that this was a city project and the developers were given a moderate price because this project was "economically important for the city government. They were eager to lower the land price."[24] As the first project in a formerly industrial area, success was important to the city.

The two project developers, De Key and Rabobank, invested fifty-fifty for the project.[25] The general approach to most design of housing units was to provide a limited variety of conservative units so the project that would be certain to sell quickly. "All the developers tell you . . . just a few people want a different apartment."[26] In this case, the decision to make a great variety of dwellings was seen as risky as the developers might be left with empty residences on their hands.

In the end, the architects and developers were vindicated.

> The market says people are very different. That story was true. Of course it's true! It's a beautiful spot and everybody wants to live there. Every house was sold. The proof is in the pudding. You don't need to build just these mainstream houses.[27]

Notes

1. Frans de Witte (Architect MVRDV), Interview, Amsterdam, May 31, 2012, 1.
2. Eric Amory (Developer, De Key), Interview, Amsterdam, May 25, 2012, 3.
3. According the density levels defined by Mozas and Per (Javier Mozas and Aurora Fernández Per, *Densidad: nueva vivienda colectiva* [*Density: New collective housing*] (Vitoria-Gasteiz: a+t Ediciones, 2006), 15).
4. Ton Schaap (Urban Planner, Municipality of Amsterdam), Interview Amsterdam, March 23, 2010, 1.
5. Our calculations of density were based on data provided by the architects (Frans de Witte, Email Communication, March 8, 2016).
6. Nathalie de Vries, "MVRDV brought different house types together with Silodam block says Nathalie de Vries." Online video clip. YouTube. Dezeen, August 10, 2015. www.youtube.com/watch?v=FmEufmB8B6E.
7. Frans de Witte, Interview, 2012, 2.
8. Ibid., 2.
9. Ibid., 7.
10. Eric Amory, Interview, 2012, 3.
11. Frans de Witte, Interview, 2012, 6.
12. Eric Amory, Interview, 2012, 1.
13. Ibid., 1.
14. Frans de Witte, Interview, 2012, 4.
15. Ibid., 6.
16. Ibid., 7.
17. Ibid., 6.
18. Ibid., 5.
19. Eric Amory, Interview, 2012, 6.
20. Frans de Witte, Interview, 2012, 7.
21. Ibid.
22. Nathalie de Vries, YouTube.
23. Frans de Witte, Interview, 2012, 5.
24. Eric Amory, Interview, 2012, 3.
25. Ibid., 3.
26. Ibid., 5.
27. Ibid., 3.

10 Design Principles for Complex Housing

Where single family housing is the ideal form, as is true in many countries, one way for dense housing to be more accepted is for its design to resemble this ideal as much as possible.[1] There are many reasons to appreciate a freestanding house, including access to light and ventilation on all four sides, lack of acoustic transfer from an adjacent unit, access to a garden or outdoor area, direct and easy access to the street, visibility and accessibility of the street and neighbors, ability to recognize one's neighbors, no sharing of sidewalks or halls, and the concomitant clarity of ownership boundaries. For many people in the Netherlands, where single-family housing is a rarity, rowhouses meet the ideal. They are associated with having direct access to the street, more than one level, a garden, and access to light and air on two sides.

Dutch complex housing, in comparison to the simple, dense model of housing introduced earlier with layers of identical housing creating a tall building, increases the proportion of the desired qualities. For example, the high requirements in the Netherlands for light and air tends to result in housing with two sides or two ends of a unit oriented to the exterior rather than just one, or with a one-sided unit that is shallow rather than deep. Another design that resembles a house is a maisonette; it may not have direct access from the ground level, but it provides the separation of private and intimate areas by level that is embodied in a traditional house. The concrete construction typical of Dutch complex housing creates very effective acoustic separation in comparison to wood or metal construction typical of housing in the United States. While there is no way to get around sharing circulation areas in dense housing, the Dutch designs restrict the numbers of units on a hall or gallery so that residents will recognize their neighbors. Because the out-of-doors is associated with the public realm and the interior is associated with the private realm, the use of exterior galleries rather than interior corridors creates a sense of circulation area as public space and encourages visibility of neighbors, especially when galleries are visible across courtyards. Finally, typically, Dutch complex housing includes an exterior space associated with the dwelling, whether garden, roof garden, gallery, balcony, or loggia. Thus we find projects that include housing with direct access from the street, courtyard, or exterior gallery; dwellings with several levels; and dwellings with large amounts of fenestration and with outdoor space.

The Dutch housing examples also avoid many of the negative features associated with density, such as appearing monolithic and overwhelming in scale; crowded and monotonous; and lacking sufficient privacy and control. This is achieved in different ways and to different degrees. The key factors are (1) massing that responds to the urban (or suburban) context; (2) building façades that reduce apparent size and monotony; (3) courtyards and open space that create communal territory within the complex; (4) connection to the neighborhood through design and provision of amenities; (5) a mix of unit tenure, target group, and type; (6) limited access to circulation spaces in support of security and social structure; (7) outdoor space associated with the unit; and (8) orientation to generous amounts of light and ventilation.

Location, Context, and Site

Context

Because of its density, complex housing is typically found in urban or suburban areas rather than rural settings, and usually on or near a main thoroughfare. Complex housing is also related to an open space. With the exception of De Beeklaan, all of the buildings studied are either on water (La Grande Cour, Vrijburcht and Silodam), a park (De Muzen and De Opgang), or a plaza (Carnisselande and De Zilvervloot), resulting in good views and high visibility. While most of De Beeklaan is located within a neighborhood context, the main tower is prominent on a very wide street, and marks the entrance to the neighborhood.

Complex housing typically faces an open space such as a plaza (Carnisselande), park (De Muzen), or water (La Grande Cour)

Dutch regional planning gives high priority to pedestrian, public transportation, and bicycle access. Easy pedestrian access to public transportation is typical, and grocery shopping and taking children to school and childcare are often done by bicycle. Complex housing therefore has easy pedestrian access to public transportation, bicycle access to shopping and activities such as childcare and schools, and includes bicycle parking and storage.

With the exception of the suburban De Muzen and Carnisselande projects, the complex housing projects build up to the edge of the lot without significant setbacks from the street.

A key site principle that was mentioned in interviews is locating functions at the street level to keep activities on the street—that is, both non-housing (commercial, civic, etc.) and rowhouses—to maximize the number of entrances. As Henk Jan Hollander proposes, "When you walk on the street you [often] only see parking garages, and I wanted to see houses, which give a better atmosphere."[2]

Box 10.1

Context: Design Principles

1. Locate the complex housing project within easy walking distance to public transportation and easy bicycling distance to grocery stores, childcare centers, schools, and other amenities.
2. Position the complex housing project overlooking as much open space as possible: a park, a plaza, water, etc.
3. Orient the project to maximize light and ventilation ("licht en lucht").
4. Build up to the street edge. Where possible, instead of providing setbacks from the street, leave space in the middle of the site for a courtyard.
5. Locate housing and other uses to maximize entrances on the street to encourage street life.

Project Functions

Complex housing projects, in addition to incorporating a variety of housing, generally include non-housing functions on the street level. When a project is located on a busy street, placing commercial or other functions at the ground level can be advantageous. Furthermore, incorporating activities that benefit the larger neighborhood can play an important role in connecting a complex to the adjacent community. While early complex housing developments such as De Muzen and Silodam had few community facilities within the building, and were located off the beaten path, later projects incorporate a significant amount. For instance, Vrijburcht incorporates workplaces, a childcare center, a theater, and a café, and De Opgang includes a church and childcare center that are utilized by people outside the project. Usually these activities are located on the ground plane, but occasionally other levels are used as well. For instance, the theater at Vrijburcht is above the street level accessible from the outdoor deck, and Carnisselande has access to the library and clinic both at the street level and the level below.

Commercial and Workspace

Commerce is the most common additional function, found with every project except De Muzen, although the amounts and types vary, from 31,000 m^2 at De Zilvervloot, which incorporates a supermarket, a restaurant, and many small shops, to only 600 m^2 at Silodam. At Vrijburcht, much of the commercial space is associated with the atelier and live-work units. Projects such as Carnisselande and Vrijburcht that include live-work units generally provide direct street access, although the live-work units at De Zilvervloot have indirect access.

Community and Civic Space

Other functions found within the projects are community space (De Muzen and Vrijburcht), educational space (childcare at De Opgang and Vrijburcht), and civic space such as theater (Vrijburcht), library and clinic (Carnisselande), and church (De Opgang).

Parking and Storage

Parking is always included in the projects, usually at 1.5 cars per unit, although in urban areas this may be reduced to 1 per unit. In the Netherlands many people use public transportation, so the number of parking places can be lower. At De Opgang, for example, developer Hen Jan Hollander provided one parking place per unit and a few extra places to accommodate other functions and was unable to sell all of the spots provided.

> I had to develop the parking because of the regulations, but these are first-time owners and 50% of the people that bought there have no car. So I spent a lot of money on a parking garage . . . [and] today I am the proud owner of 12–13 parking places.[3]

At suburban sites parking is predominantly on grade as at De Muzen. Carnisselande mixes on-grade parking with a small proportion under the building. The remainder of the projects have parking below grade or under the building in some form, whether fully below grade as La Grande Cour, or partially as De Opgang. De Beeklaan parking is all on-grade either under the building (flats and rowhouses), in a parking area (flats), or underneath a deck that serves as outdoor space for a rowhouse. At Vrijburcht, the L-shaped parking is only under one part of the block to allow for tree growth and water collection.

In addition to parking, all projects provide storage areas, either associated with the below-grade parking or within the unit. Bicycle parking is also included, sometimes in a designated area and sometimes in a shed or storage area. To prevent residents from parking their bicycles on the sidewalk, bicycle areas should have access to the sidewalk that is direct as possible.

A variety of functions are incorporated in complex housing projects: commercial (De Zilvervloot), civic such as church and childcare center (De Opgang), theater and café (Vrijburcht), and parking (La Grande Cour)

Box 10.2

Project Functions: Design Principles

1. Place parking below grade where possible, providing a maximum of one parking unit per dwelling.
2. Use the street level for rowhousing, commercial space, and other non-housing purposes.
3. Include general storage as well as secure bicycle parking with easy access to the street.

Site

The eight projects employ very different strategies to fit their context, in some cases accepting the role of background building, like Vrijburcht and De Beeklaan, and in others featuring their monumentality. The varying contexts and sizes of the projects offer a great range of approaches to site design that do not necessarily reflect the actual number of units in the project.

Size and Height

The number of dwellings in the projects range from 52 to 253, with Vrijburcht and De Opgang at the low end, being under 100 units; De Beeklaan and De Zilvervloot at the low middle (under 150); and De Zilvervloot, Silodam, and De Muzen at the upper middle (between 150 and 200 units). The outlier is La Grande Cour at 253 units, with 82 more units than Silodam. Because both suburban and urban projects are included, the floor area ratios and building heights do not necessarily parallel the number of units. Furthermore, as we have seen, the variation in clients' needs and desires generated different design solutions (see Table 10.1).

Net site density ranges from 77 to 604 units per hectare (31 to 241 units per acre). The footprint density, not directly comparable to net site in terms of numbers, shows a smaller range that may better reflect the experience of the housing density. These figures show Silodam as still the most dense project, but followed by De Muzen when its large site is taken out as a factor. The remaining sites are in similar relationship as before, although De Beeklaan becomes less dense instead of more dense than De Zilvervloot and Carnisselande. Vrijburcht, whose footprint is identical to the site, trails significantly behind the others. The advantage of the footprint calculation is that it allows a designer to know the density implications of a given design footprint without having to do additional calculations.

Floor area ratios include the other functions in the project such as commercial, civic, or parking. The highest are Silodam (7.5) and La Grande Cour (6.5), demonstrating their tight site conditions. De Zilvervloot's significant parking and commercial levels contribute to its relatively high figure of 4.07, not much greater than De Opgang's 3.41. The fourth grouping finds Carnisselande, Vrijburcht, and De Beeklaan at the lower end of the cohort, and De Muzen trails significantly at the end with a .76 FAR.

With the exception of Silodam at ten stories, the projects do not have a uniform height. Most of the projects combine midrise (four to ten stories) with low-rise (one to three stories) and/or towers (twelve to eighteen stories). The main part of complex housing projects tends to be six to eight stories. Only Vrijburcht, Carnisselande, and De Muzen have sections substantially lower, each with a large portion three

Table 10.1 Project size

Size of Projects

	# Units	Density 1 Net Site units/ha (units/acre)	Density 2 Footprint (units/hect)	Gross Floor Area (m2)	Floor Area Ratio (FAR)	Maximum # Stories
De Muzen	167	77 (31)	340	16,500	0.76	4
Vrijburcht	52	118 (48)	118	11,400	2.59	7
De Zilvervloot	130	170 (69)	203	31,100	4.07	7
Carnisselande	171	193 (78)	211	25,990	2.93	18
De Beeklaan	105	197 (80)	182	13,417	2.52	6
De Opgang	79	242 (98)	249	11,160	3.41	6
La Grande Cour	253	351 (142)	297	46,800	6.50	10
Silodam	157	604 (244)	440	19,500	7.50	11

Silodam is designed as a single volume, De Opgang is expressed as two parts, De Zilvervloot is a perimeter block composed of many individual forms, whereas De Beeklaan, a traditional design, comprises several separate blocks

stories in height. The tallest sections that are not towers are four stories at Vrijburcht; range to six stories at De Opgang and De Muzen; range from two to eight at De Beeklaan; and are eight stories at La Grande Cour's and Carnisselande's main blocks.

The towers generally contrast to lower housing. Four of De Beeklaan's five blocks on the main street are four stories (rowhouses two and three stories), with the tower block stepping up from five to six and seven stories tall. De Zilvervloot's lower sections range from four to eight stories, with a twelve-story tower. While La Grande Cour's rowhouses are only two stories in height, the main block is eight stories and the periscopes are eleven. Carnisselande's tower is eighteen stories tall.

Massing

Complex housing displays a variety of massing, falling on a continuum between being a single slab and being composed of many volumes of differing shape. Silodam is a pure example of a single slab, with De Muzen also being a single slab with the added complexity of having three lengthwise sections and an appendage to the north that creates the interior courtyard. Relatively straightforward perimeter blocks are formed by Vrijburcht, and De Opgang, while De Zilvervloot and La Grande Cour have more articulated blocks. Designed as

De Muzen has a tall central section that steps down to the two 3-story lower ends, and is intersected by a bicycle and pedestrian path on each side

a set of partial blocks, De Beeklaan forms fingers that fit into existing independent blocks. Conversely, the slabs at Carnisselande create fingers that form new blocks.

Massing can be used to reduce the apparent scale of the projects. By breaking down the mass into parts, what might be perceived as a large monolithic building becomes several smaller elements. So for example, De Muzen, from the south an extraordinarily long block that encloses a park, has a reduced apparent scale by breaking down the long building mass into a middle section seven stories tall, and two end sections, as well as the four tulip-shaped blocks that enclose the interior courtyard at the community side.

Slab

From examining the floor plans (see illustration, next page), we can see that two of the projects are primarily composed as slabs (Silodam and De Muzen). Five of the other projects (Vrijburcht, De Beeklaan, De Opgang, De Zilvervloot, and la Grande Cour) are composed of slabs that form a perimeter block with one or more courtyards in the middle. And one building, Carnisselande, is a kind of hybrid, having a main slab, accented by a tower, from which fingers frame two perimeter blocks surrounding courtyards.

Perimeter Block

Vrijburcht, De Beeklaan, De Zilvervloot, De Opgang, and La Grande Cour are perimeter block buildings, like most complex housing. The exceptions are Silodam, which is clearly a single slab; De Muzen, which is a long slab with a second parallel slab in the middle that creates the interior courtyard between; and Carnisselande, which has courtyards but is not walled on all four sides. These three exceptions each respond to a particular context. Silodam, being in the river, is massed more like a boat than a city block. In the original design for that area of suburban Almere, De Muzen was envisioned as a semi-circular building at the end of the adjacent park. The architects, responding to neighbors' objections, made the building a flatter arc so that it would be less intrusive in the neighborhood. Carnisselande, on a triangular site, also responds to its adjacent structures, being a wall on the side of the town square and open to the housing behind. It could be said to form fingers rather than a clear perimeter block (De Beeklaan forms fingers too, but these fit into a perimeter block). Five of the eight projects exist within the urban block fabric of the city, and respond to their context by forming their housing into a perimeter block.

Tower

Three buildings have towers: those at Carnisselande and De Zilvervloot are eighteen and twelve stories, respectively, while the tower at De Beeklaan is a modest seven stories. At eleven stories in height, the periscope buildings could be considered to be towers, which they appear to be when seen from the courtyard. However, from the exterior they appear as cantilevers. The towers and periscopes all have a kind of cap. In the case of Carnisselande, it is simply a wide roof cantilever. De Zilvervloot is capped by the two 2-story luxury units that have a silver cladding and punched windows, in contrast to the window walls on the floors below. De Beeklaan is topped with a six-sided pitched roof and cupola.

The floor plans demonstrate the slab (De Muzen and Silodam), perimeter block (Vrijburcht, De Zilvervloot, La Grande Cour, De Opgang), and fingers (Carnisselande, De Beeklaan) arrangements

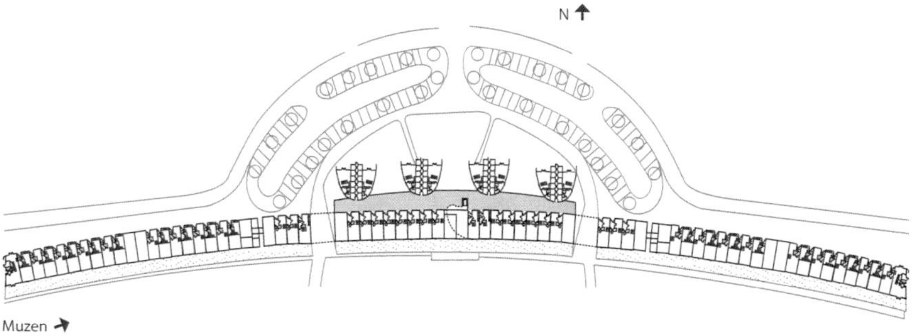

De Muzen

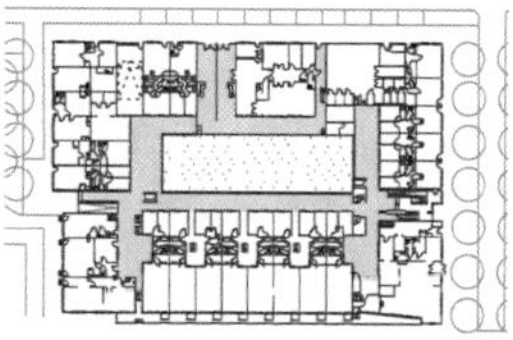

Vrijburcht

De Zilvervloot

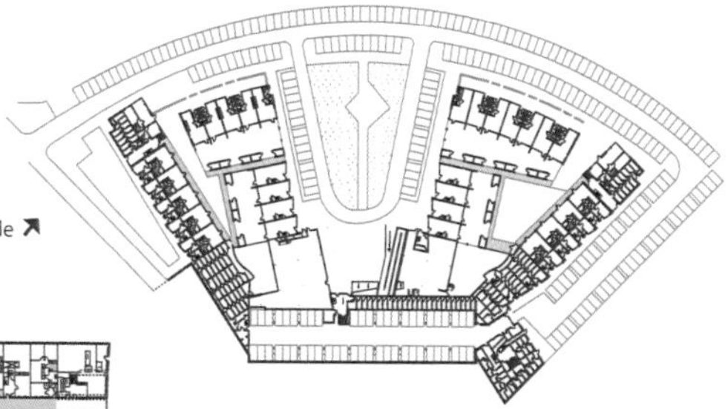

Carnisselande

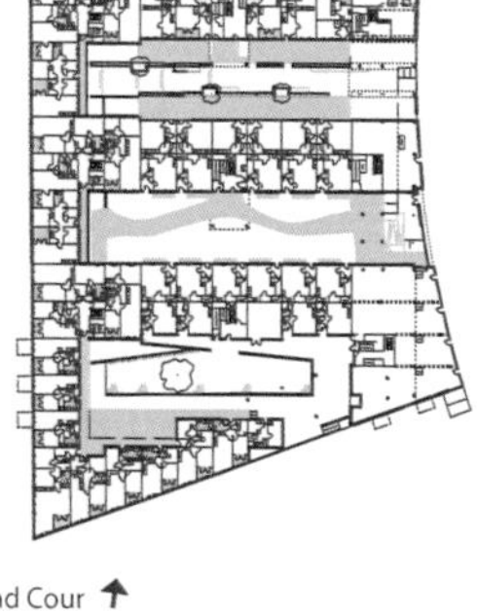

La Grand Cour

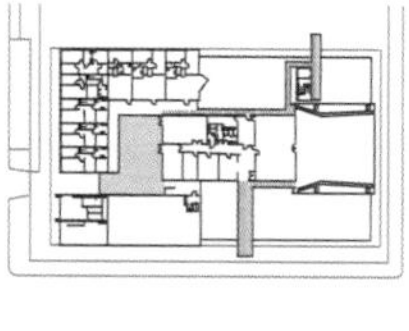

De Opgang

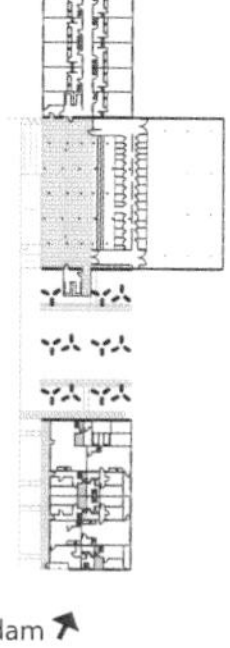

Silodam

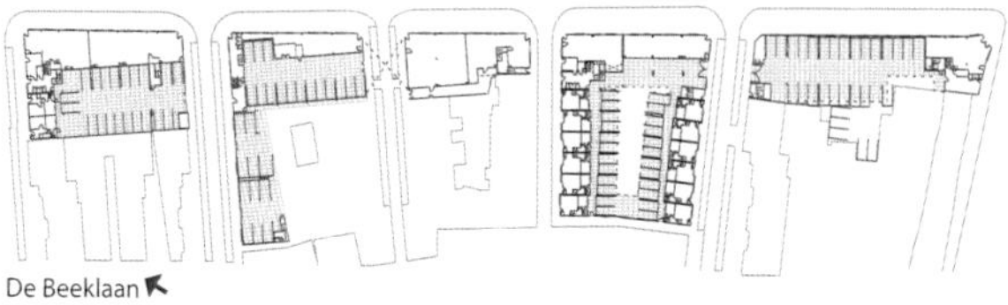

De Beeklaan

Bridges, gates, and windows in the building mass allow light and air to penetrate the façade at La Grande Cour

Bridge, Gate, and Window

Many of the projects use bridges to create entry gates, or urban scale windows that allow light into the interiors of courtyards. Perhaps the most dramatic bridges are at La Grande Cour, where two levels of gallery-accessed units are suspended between vestibule apartment buildings on the east side. In other projects entries are created by housing bridges, such as the bike and pedestrian paths through De Muzen, or the gate formed by the link between the plaza and library and clinic at Carnisselande. La Grande Cour also has significant wall openings: one in the perimeter wall on the south side, and others between the courtyards, where the periscope buildings bridge the two-story rowhouses to create four- or five-story openings.

Box 10.3

Site: Design Principles

1. Keep the main part of the building three to eight stories tall.
2. Use varying heights to express the different parts of the complex and/or to allow sun into a courtyard.
3. Reduce the apparent scale by:
   a. Creating a series of small masses rather than one large one.
   b. Creating bridges to make openings in the building wall and to allow light to pass.
4. When including a tower, limit its height to eight to eighteen stories, and top it off with a cap.
5. Consider the heights of the adjacent buildings and use massing to blend with the context.
6. Consider using a narrow building footprint to allow space for a courtyard.

Articulation of Form

Dense buildings are more acceptable to the public if they minimize their apparent size. In addition to and often emphasizing massing, project designers can employ a variety of articulation strategies to reduce the scale of a building, such as setbacks (being raised at the base, having a setback at the top), overhangs or cantilevers (used to differentiate large or small masses; to break up smooth façades into parts), openings (for entries, to allow light in) and negative space (balconies, loggias, and openings), fenestration (can blend with the wall or distract the eye from the wall, can differentiate one building part from another), material, and color. The ways the designers use such elements contribute to the building's character and identity. In each case, the unique blend of setbacks, cantilevers, fenestration, material, and/or color creates a memorable building.

Setbacks

A building raised at the base or set back at the top appears lighter and thus less large. This approach is used to indicate the commercial level both at the perimeter of La Grande Cour and the front façade of Carnisselande. La Grande Cour also employs this tactic by having a smooth skin to express the building volume and creating a shadow line where the commercial functions are located on the east, south and west sides. On Carnisselande's south façade, the negative space at the base complements the other articulations such as variations in the tower material, and the white cantilevered block of units with larger floor plates.

When setbacks are used at the roofline, like the top of De Muzen, which is offset from both the front and sides, the lower part of the building appears to be the top rather than the receding top layer, thus making the building seem one story lower. The effect is especially pronounced in the photograph taken in 2007 when the wood siding on top layer was a much darker color than the brick below (see photo lower right).

Cantilevers, Balconies, and Fenestration

At De Opgang, the overhang highlights the independence of the church from the housing. The cantilevered building masses of the periscope buildings are used at La Grande Cour to create additional units without adding a uniform layer across the entire building. Similar to setbacks, because the building is taller at the middle of the building rather than at the edges, cantilevers make the building's overall height less obvious. Additionally, the projections create a sense of suspension that is very exciting to the observer.

Balconies and windows tease the eye on the south façade of De Muzen (see photo below, right). To create a playful pattern, the interior unit plans are arranged so that the living rooms and their associated balconies do not line up regularly. A variety of windows punctuates the façade. At De Zilvervloot, the architect layered units so that balconies and window variation create contrasting vertical patterns on the building façade. At La Grande Cour, we find a more restrained approach to articulation (see photo page 215, left). On the east, windy side, while color is used to differentiate the windows over the center entrance, the use of loggias instead of balconies for exterior access creates a smoother façade that foregrounds the periscope cantilevers. The regular fenestration and loggias used in each block at De

To reduce apparent size, Carnisselande uses varied building heights, along with overhangs at midsection and tower, and setbacks at the base of the building

Balconies and setbacks at De Muzen give human scale to what might otherwise appear as a monolithic wall

Uniform fenestration and the use of loggias on the north side of La Grande Cour emphasize the building surface and overall volume

Beeklaan create horizontal lines that de-emphasize the building height and connect the different blocks.

Fenestration

Fenestration can be used to emphasize the opening, as in the center of the east façade at La Grande Cour, or to emphasize the surrounding wall, as the loggias on the sides of the same façade (see photo above). In the first case, the windows are outlined in white. In the second case, the loggia frames match the surrounding materials in color. But here the size of the openings is relatively similar, the color differences are subtle, and the wall still reads as a smooth plane.

In contrast, at De Zilvervloot we find that the different types of interior spaces are expressed on the façade with noticeably diverse fenestration (see photo, upper right). Each area of the building has a different type of window, placement, and color. The skip-stop dwellings in the red block have an irregular pattern of opening, framed in white to contrast with the red wall that calls attention to itself. The white wall and repeated pattern of white-framed windows and balconies of the vestibule housing create a background section that emphasizes the verticality of that part of the complex.

Color and materials are used to differentiate masses and to express functional differences at De Zilvervloot

Material and Color

In contrast to the more sober approach taken at La Grande Cour, De Beeklaan, Carnisselande, and Vrijburcht, which tend to read as unified structures because of their consistent materials, colors, and patterns, other buildings use color and materials to express various degrees of unity. Perhaps the strongest differentiation is found at De Zilvervloot, where different colors and materials are used in each building mass to create a sense of a multitude of buildings rather than a single structure. Color reinforces horizontality at both De Muzen, with the contrast of the coral façade to the darker wood crown, and at De Opgang, where the green housing superstructure is a counterpoint to the red brick base. At Vrijburcht slight color differences identify functional areas. At Carnisselande the building reads as a whole but color and material—here white concrete—is used to accentuate building elements against the main salmon colored slab.

Box 10.4

Articulation: Design Principles

1. Use setbacks at the bottom of the structure to reduce the tie to the ground and make the building float.
2. Use setbacks and cantilevers at the top to minimize the overall building height.
3. Use setbacks, cantilevers, fenestration, and balconies in the building wall to create shadow lines that reduce the amount of smooth surface.
4. Consider the relation between the arrangement of plans and the location of fenestration and balconies. If a regular pattern is desired, line up the plans; if a more playful pattern is desired, consider altering plans on different floors so that windows and balconies have a less predictable arrangement.
5. Employ contrasting material and color to express the building parts and to add liveliness or use more uniform materials and colors to create a more subtle or background presence.

Courtyard

All of the complex housing projects have courtyards with the exception of Silodam. The courtyards serve a variety of functions (access, community, garden, visual, play, lightwell, commerce, and parking), often several simultaneously, each one being unique.

The visual, garden, and social courtyards are designed to be sunlit much of the day. The structures on the south side are generally shorter and often elongated north to south to allow sunlight to enter the courtyards. Sometimes, as at La Grande Cour, Carnisselande, and De Zilvervloot, the taller buildings have a small footprint so the shadow moves during the day. Additionally, there may be openings in the building walls to allow light into the courts.

The materials used in the complex housing courtyards vary according to use, from all hard surfaces in access courtyards to large amounts of vegetation in social, garden, and visual courtyards. It is possible to include trees, which greatly enhance the character of a court, when there is no underground parking garage, when the garage is limited to part of the site as at Vrijburcht, or when a deep and wide box is created for the roots. While color can enliven a courtyard, as in the middle court at La Grande Cour, combining hard or man-made surfaces with plant materials creates a more inviting environment, such as in the north and south courtyards.

Access

While many projects have their primary access directly from the street, others enter through a courtyard. At La Grande Cour, the three courtyards have distinctly different designs, although they all provide access to the units from the canal quay to the east and/or the commercial street on the west. Such courtyards are primarily entrances and circulation spaces, as they are points of access to all of the dwellings, indirectly through lobbies to flats and maisonettes and directly to the back-to-back rowhouse social housing. Other courtyards that serve as pedestrian entry access are those at Vrijburcht and De Muzen and the residential courtyard at De Zilvervloot.

An access courtyard at La Grande Cour (north courtyard looking east)

The atrium at De Muzen is a garden courtyard

The community courtyard at Vrijburcht

Community

Vrijburcht and De Muzen are the most social courtyards because they not only serve as the main circulation point for the project, but also incorporate furniture and green space to encourage use. Vrijburcht's small number of units and its low height keep the courtyard intimate and sunny. Similarly, the atrium at De Muzen is an effective community courtyard because it houses the entrances of many of the dwellings. Although designed as community spaces and open to all of the residents, the community courtyards at De Zilvervloot and Carnisselande are used on a daily basis by only a few and are indirectly accessible to the others. The highly accessible courtyards at La Grande Cour are designed for moving through rather than for social interaction, so may not be particularly effective as community courtyards.

Garden

Many courtyards include vegetation, but what we are calling a garden courtyard is used by some or all residents. This is differentiated from courtyards that are inaccessible and only serve as a view for residents, as a visual courtyard or as a lightwell courtyard. In this definition, the primarily social courtyards at Vrijburcht, De Muzen, De Zilvervloot, and Carnisselande can also be called garden courtyards. Designed by a landscape

A visual courtyard at Carnisselande that also serves as a community courtyard

Two of the three courtyards at De Opgang are lightwell courts

The third, play courtyard at De Opgang

architect, the interior atrium courtyard at De Muzen is created to be the heart of the project. While its most obvious purpose may be as an access route to the surrounding units, it also serves as a year-round, plant-filled acoustically quiet garden space. The court at Vrijburcht also serves as a rain garden that collects, treats, and reuses the water. At Carnisselande the courtyards serve as gardens for the surrounding dwellings and include a community green space for the complex.

Visual

Although among this group of projects there are no exclusively visual courtyards, several serve this purpose in addition to others. For example, the atrium at De Muzen, as well as the courtyards at Vrijburcht and Carnisselande, offer views for the units that overlook them.

Lightwell

The upper two courtyards at De Opgang, being inaccessible and one without a landscape design, are essentially lightwells. With careful landscaping, the middle court has potential to become a visual courtyard, like the green rooftop court.

Play

The courtyards at Vrijburcht and De Opgang are designed for children's play. At Vrijburcht the courtyard serves both the childcare center and the residents. The large amount of vegetation mutes the sound of play. At De Opgang, the lower courtyard, surfaced with artificial turf and incorporating play equipment, serves as a play area for the childcare center that can be noisy for neighbors during the day.

Parking

Unique to the group of projects studied, De Beeklaan employs courtyards to provide affordable parking within

The parking courtyard at De Beeklaan

De Zilvervloot's commercial courtyard

the city, serving as alleys that allow automobile and bicycle parking, as well as service for the larger buildings. Reached through gates or gaps in the building wall, the parking is located at ground level underneath the apartment structures or the decks of the rowhouses.

Commerce

The only project with a commercial courtyard is De Zilvervloot. Serving as the main access to the two supermarkets and a restaurant, it is actively used. Because the businesses serve the whole neighborhood, De Zilvervloot and the adjacent public plaza have become its heart.[4]

Box 10.5

Courtyard: Design Principles

1. Combine housing volumes to create an outdoor open space.
2. Orient the open space and vary the height of the housing so that sun will light the courtyard for all or most of the day.
3. Consider creating openings or bridges in the building wall to create entry gates or windows into the space.
4. To create an inviting environment, combine hard and man-made materials with vegetation.
5. Select a use for the courtyard that is advantageous to the dwellings and other activities that open onto it; consider whether it should, and how it might, serve the adjacent neighborhood.
6. If a courtyard is intended for shared use, it needs to be easily accessible, sunny, and furnished for sitting and conversing.
7. Attend to sound transmission. Courtyards with all hard surfaces can be very reverberant. Vegetation can mute sound levels.
8. Where possible, use the courtyard to gather and treat water.
9. When designing an underground parking area, consider leaving parts of the site free of construction to allow space for the roots of trees in the courtyards.
10. Courtyards can contribute socially as well as visually to the project. Employ a landscape architect to design the courtyard so that it is both useful and attractive.

Housing and Unit Type

Perhaps the most characteristic aspect of Dutch complex housing is its composition of many different housing and unit types, and the use of many types of access. As

Table 10.2 Form dwelling types

Projects	Form Dwelling Types											
	Multi-story Units w/Various Access						% Flats w/Various Access					ALL
	TOTAL ROWHOUSES	Skip-stop maisonette	Gallery maisonette	Vestibule maisonette	Corridor maisonette	TOTAL MAISONETTES	Gallery flat	Corridor flat	Vesti-bule flat	Direct-access flat	TOTAL FLATS	
De Muzen	**39%**		11%			**11%**	50%				**50%**	100%
Vrijburcht	**19%**		35%			**35%**	2%		44%		**46%**	100%
De Zilvervloot	**2%**	6%	11%	3%		**20%**	18%	24%	34%	2%	**78%**	100%
Carnisselande	**19%**			20%		**20%**	59%				**71%**	100%
De Beeklaan	**13%**					**0%**	82%		5%		**87%**	100%
De Opgang	**10%**					**0%**	90%				**90%**	100%
La Grande Cour	**11%**	14%				**14%**	26%	11%	38%		**75%**	100%
Silodam	**0%**	19%	6%		19%	**44%**	10%	41%		5%	**56%**	100%

Table 10.3 Purpose dwelling types and totals

Projects	Purpose Dwelling Types							Totals (Form + Purpose)		
	Penthouse	Lightwell	Live-work	Group Home	Assisted Living	Guest Units	TOTAL % OTHER HOUSING	# Housing Units	# Housing Types	# Unit Types
De Muzen	6%				60%	1%	**7%**	**167**	**6**	**11**
Vrijburcht	4%	44%	19%	8%		2%	**77%**	**52**	**8**	**14**
De Zilvervloot	4%			7%			**13%**	**130**	**5**	**29**
Carnisselande			8%	8%			**16%**	**145**	**5**	**12**
De Beeklaan	1%				31%		**31%**	**105**	**5**	**9**
De Opgang							**0%**	**79**	**2**	**8**
La Grande Cour	10%						**10%**	**253**	**5**	**39**
Silodam		5%					**10%**	**157**	**4**	**17**

mentioned earlier, simple dense housing typically includes one housing type and as few as one to three unit types. All of the projects discussed here have between seven and twenty-eight unit types, and most have three or more housing types, the exception being De Opgang with two housing types and eight unit types. The projects also typically incorporate both flats and two or more story dwellings in the form of rowhouses and/or maisonettes.

As earlier, in this analysis we discuss both the housing and unit types with an emphasis on housing type. Although access is addressed later as a typology in and of itself, the discussion of housing and unit types must include type of access since this is a factor in characterizing different types of housing. After classifying all of the housing by its form (the number of levels and mode of access—rowhouse, maisonette, flat), we identify special housing by its purpose and use (penthouse, lightwell units, group home, assisted living, and guest flat), and then categorize unit type. Purpose-based types may employ one or more form-based housing type in their construction. To illustrate the great variety of unit types, the associated chart uses a similar approach but combines these different categories (showing more variation) (see Tables 10.2 and 10.3).

Housing Type

Rowhouse

Rowhouses are found at all of the projects except the densest, Silodam. The several Silodam units that enter directly from the dock rather than through the main entrance are flats. Rowhouses are entered directly from the street at De Opgang, Carnisselande, De Beeklaan, and De Muzen, but indirectly through a courtyard at Vrijburcht, La Grande Cour, and De Zilvervloot (where the courtyard is one level above grade, and requires indirect access though an elevator core). At Carnisselande, De Beeklaan, and De Zilvervloot the rowhouses are mostly three stories tall, while at De Opgang, De Muzen, and Vrijburcht they are all two stories. While the majority of the rowhouses are embedded in the housing block, at De Beeklaan and Carnisselande there is a more traditional arrangement, in a rowhouse block with no units above.

Embedded rowhouse units at De Opgang and De Muzen and freestanding rowhouses at De Beeklaan

Most of the rowhouses in complex housing have a floor-through design with orientation to light and air on both ends of the unit. In some cases, garden space or patio space is provided at the front of the unit (De Zilvervloot and De Beeklaan), but more often either a garden space is provided on the ground floor at the rear (Carnisselande and De Muzen), or a deck or balcony is provided at a level above the ground (De Beeklaan, De Zilvervloot, La Grande Cour, Vrijburcht, and De Opgang). However, at La Grande Cour the majority of rowhouses are designed back-to-back with a single orientation, and of these only one-third—those in the north courtyard—have access to a viable outdoor space. At Vrijburcht the rowhouses double as live-work atelier

Skip-stop maisonettes at Silodam, in De Brug at De Zilvervloot, and in the north periscope building at La Grande Cour

dwellings with both direct street access for work, and deck-level residential access. At Carnisselande some of the rowhouses also serve as live/work units.

Maisonette

Maisonettes are found in many of the complex housing projects. Their design varies greatly both within and across projects. There are no maisonettes at De Opgang, Carnisselande, and De Beeklaan, and a small number at De Muzen. At Silodam, Vrijburcht, De Zilvervloot, and La Grande Cour, maisonettes take many forms. Only at Vrijburcht are the majority of units maisonettes or rowhouses, although they constitute 50% of the units at De Muzen and 44% at Silodam.

The greatest variation in maisonette design is at Silodam. The most unusual are the twelve cross-over dwellings (also sometimes called X design), in which the two levels are not on top of each other but over an adjacent unit and facing different sides of the building. This provides cross-ventilation and variety of orientation. Another style of maisonette at Silodam has a floor-through design that goes across the whole building on the upper or lower level. Finally, half of the Silodam maisonette units are

organized in a back-to-back format along a single-loaded corridor with view and ventilation to only one side.

At De Zilvervloot there is also a great range of maisonettes, although they make up only about one-sixth of the units. For example, the eight paired skip-stop maisonettes that form the Brug (or bridge) have access on both east and west, enter on the sixth level, and either go down one level or go up two.

At La Grande Cour, the maisonettes that represent about one-sixth of the units are all skip-stop units. Perhaps the most unusual are those in the north and middle periscope buildings. Fourteen of these enter at the ninth floor and go up two more stories to a roof garden at the top, and nine enter at the ninth floor, but go down to an east-facing balcony. Also of interest are the many skip-stop maisonettes at La Grande Cour that take up three floors of the east wall. These are entered at the third level and either go down one floor or up a floor to their main living area and loggia on the east or balcony to the west.

The maisonettes at Vrijburcht are located on the northeast and southwest sides of the project and are reached through the elevators and wide access decks on the two upper levels. Several maisonettes on the northeast side of the project have living space on levels 2 and 3, with their front door on level 1 and access via stairs.

Flat

Every complex housing project contains flats, perhaps the most efficient type of dwelling for dense housing; most projects exhibit a large variety of flat types. Double-loaded corridor access for flats, commonly used to create dense housing in other places, has more restricted use in complex housing. Only three projects employ this approach: De Zilvervloot (24%), Silodam (41%), and to a lesser extent La Grande Cour (11%). In each of these cases, the corridor length is restricted to a small number of flats by fire-code doors that delimit a section of the building.

Gallery-accessed flats are the most common housing type in complex housing. They are typically used for social housing and moderate-income housing. At Opgang, Carnisselande, De Beeklaan, and De Muzen, they make up 50% or more of the housing. This floor-through design allows light on both sides of the unit, and results in narrow blocks that are ideal for forming courtyard walls. When gallery-accessed corridors occur at every level, they tend to look monotonous from the exterior, and are thus usually placed on the back of the building facing a courtyard, as at De Muzen where most of the gallery-accessed flats face the interior courtyard. However, at Silodam there is no back side to the building, so the dozen or so gallery-accessed flats are expressed on the exterior as partial openings in the building wall. They do not appear monotonous, however, because the mix of types assures that no group of building type is longer than one-fourth of the building, and the gallery apartments are only on one or two floors between other unit types.

A high proportion of the flats at De Zilvervloot (34%) and La Grande Cour (38%) are reached through an elevator vestibule that limits the number of dwellings on a given level. This type of housing arrangement usually serves middle- and high-income units in complex housing. At Vrijburcht the 44% of flats that are vestibule access enter through staircases that typically serve six dwellings designed for low-income residents. These are discussed later as lightwell flats.

Special Housing Categories

Because diversity is such an important aspect of complex housing, it is interesting to note that several of the projects include unusual housing.

Live-Work Unit

Live-work units are found at Carnisselande, Vrijburcht, and De Zilvervloot. There is no standard design for these units; they only require that there be an extra living space for work and access from the street, ideally direct. However, at De Zilvervloot the access to the unit requires using an elevator, and at Carnisselande some of the units have the workspace upstairs on a third level. At both Vrijburcht and Carnisselande, live-work units enter from the street. Vrijburcht has two different types of live-work space, one of which is called an atelier. The difference between the live-work

Live-work units at Carnisselande for which the workplace is on the top floor

and atelier units at Vrijburcht is that the atelier units enter a half-level below the lowest living area and incorporate a one-and-a-half-story space with lighting from the tall exterior wall, while the others have the workspace separate from but at the same level as the lowest, bedroom level.

Group Home and Assisted Living

A good aspiration for a larger housing project is to include people with disabilities and the elderly in the proportion to which they exist in the larger society. Two of the complex housing projects include group homes: Carnisselande includes housing for people with developmental disabilities, and Vrijburcht for those with schizophrenia. De Beeklaan includes an assisted living program for seniors, and De Muzen was designed as housing for the elderly, although people of other ages also live there. Although there is now a change in ideas with housing people with disabilities in independent dwellings as opposed to group homes, housing design should take into account designs that will serve these groups.

Penthouse

Additional special housing has to do with developing particular prototypes. La Grande Cour, De Zilvervloot, and De Beeklaan all offer penthouses to high- or middle-income buyers or renters. Such units have roof gardens and may be larger in size than typical units. The dwellings may be similar to each other as at La Grande Cour, where a full 10% of the units are penthouses, or they may be unique as at De Zilvervloot (4% of units). At De Beeklaan there is only one such unit, and at Vrijburcht there are three, but several units have the capability to add roof terraces at a later date, and others could be combined to make a larger dwelling.

Unit Type

Back-to-Back

Back-to-back units are found in twenty-one rowhouses at La Grande Cour and eleven maisonettes at Silodam that enter from a double-loaded corridor on level 4 and go down to the bedrooms on the level below. At Silodam, because the building is oriented east and west with sun and views on both sides, there is little problem with having a single orientation other than lack of cross-ventilation. However, at La Grande Cour, the back-to-back rowhouses are deep in the courtyards and are oriented either south or north, which means that nine of the thirty-six units appear to receive no sun at all.

Cross-over

The only cross-over units in the study are thirty-nine dwellings in two sets of units at Silodam. The twenty-five maisonette, skip-stop units enter from double-loaded corridors on levels 5 and 7 and go up one floor across the building, so that they have kitchen, dining, and living areas on the lower level facing one direction and bedrooms on the level above facing the opposite direction, thus providing two different orientations and cross-ventilation. The second set of twelve cross-over dwellings have access to one side of the complex, crossing sideways over adjacent units that face the same direction.

Skip-Stop

As discussed under maisonettes, skip-stop units are found in three projects: De Zilvervloot, La Grande Cour, and Silodam. The De Zilvervloot skip-stop units form Le Brug, the entry bridge into the commercial courtyard. At La Grande Cour, fifteen skip-stop units are in the long part of the north and middle periscopes and nine are in the main block on the east side. At Silodam, fifty-four units, more than a third of the total, are skip-stop, including the back-to-back and cross-over units discussed earlier, and an additional eleven units that enter on level 9.

Lightwell and Atrium

Another type of unit that is noteworthy is the lightwell unit. Such units are found at Vrijburcht and at Silodam. At Vrijburcht the 44% of the housing that are lightwell flats are long and thin and stacked up for three levels on the south side of the project. The skinny flats are designed to maximize the number of units with a view and balcony toward the water. The covered atrium staircase access in the middle provides light and ventilation to the center of each dwelling. The eight lightwell flats (5%) on the top floor of Silodam have an outdoor atrium in the middle of the unit and are reached through a double-loaded corridor.

Box 10.6

Housing, Unit Type and Unit Design: Design Principles

1. When designing a housing block, use a variety of housing to encourage a variety of inhabitants and create a less monolithic structure.
2. To create activity on the street and provide good housing for families, put rowhouses on the ground level.
3. For corner positions and towers, employ vestibule access to apartments and/or maisonettes.
4. Ensure that some flats are on the ground floor or accessible by elevator for accessibility.
5. Consider using maisonettes as they resemble houses in their arrangement, separating living and sleeping quarters by level.
6. For street walls, consider using gallery flat, gallery maisonette, or skip-stop elevator housing. These can also be used to create bridges over gateways to a courtyard.
7. To create a socially balanced community, include housing for single people, housing for families, housing for the elderly, and housing for people with disabilities.
8. Include housing for low-, moderate-, and upper-income levels.
9. For flexibility and stability, combine rental housing with housing for purchase.
10. Provide sunlight, two orientations, cross-ventilation and outdoor space for every dwelling unit.

Access

Perhaps variety of access is what most differentiates complex housing from simple dense housing. Placement of elevators and stairs, a primary organizational strategy for any housing, determines how many units will share the circulation spaces, and how many elevators will be required. In complex housing, where several different entrance strategies coexist, their relation creates different possibilities for courtyards, gates, bridges, and so forth. Furthermore, the circulation system tends to generate the social structure, so an investment in a good system will pay social dividends throughout the life of the project.

In the United States most apartment buildings are designed with double-loaded corridors, although vestibule designs are sometimes found in luxury buildings where there is a small number of large units on a single floor. Direct access typical of rowhouses is mostly restricted to stand-alone buildings and very rarely combined with apartments in a larger block. The use of gallery access to housing units further differentiates practices in the Netherlands and other European countries where it is common, from those in the United States where it is unusual, especially in states with a severe winter climate. But as we discussed earlier, in the United States, there is rarely more than one type of access in a given structure, and the type of access typically defines the character of the building. Table 10.4 shows that all of the complex housing projects have at least two types of horizontal access, and most have three or four.

In complex housing, circulation spaces are used as a collective structuring device to create socially viable projects. Careful consideration is given to how many dwellings share an elevator, a gallery, a corridor, or a vestibule. The quality of the space is also considered important with these spaces being colorful and having natural light. As stated by MVRDV's Frans de Witte in regard to Silodam, "We also wanted to create a routing inside the building that is interesting to go [through]."[5]

Elevator and Stair

The number of units per elevator drives decision-making in multi-family housing. Three projects have only two elevators—De Opgang, Vrijburcht, and De Muzen—although each of these projects has ground-accessed rowhouses. De Muzen has the most units per elevator at fifty-two, with Silodam a close second at fifty, De Opgang at thirty-six, and Vrijburcht at sixteen. These differences reflect very different orientations to economics and social issues. For Vrijburcht and Silodam, social connection was a primary motivating

Table 10.4 Access Types

Projects	Horizontal Access				Vertical Access		
	Direct	Gallery	Vestibule	Corridor	Elevators	Units w/ Elevator Access	Units per Elevator
De Muzen	39%	61%			2	104	52
Vrijburcht	19%	37%	44%		2	31	16
De Zilvervloot	4%	29%	37%	30%	5	129	26
Carnisselande	19%	71%	10%		4	143	36
De Beeklaan	13%	81%	5%		4	90	23
De Opgang	10%	90%			2	72	36
La Grande Cour	11%	40%	38%	11%	9	226	25
Silodam	4%	16%		79%	3	149	50

The staircases at Silodam are different colors

factor; for De Muzen and De Opgang, economics had a critical effect on the design. Henk Jan Hollander, the developer at De Opgang, discusses the economics of elevators and the additional importance of maintenance:

> Having two elevators is better than 4 or 6. If one is broken, people can easily walk around the site, to use the elevators. Elevators are expensive not only in the beginning but also in maintenance. You want to keep the service costs as low as possible.[6]

Entrances to almost all of the complex housing dwellings are at ground level or reached by elevator, with two exceptions that use stairs. At De Muzen, the tulip maisonette units at De Muzen enter from the interior courtyard via stairs, and at Vrijburcht all of the lightwell units (half of the units overall) are reached via stairs.

Ground

Ground access can be direct from the street or indirect via a courtyard. Because residents tend to see and meet each other at entries, the choice whether to activate the street or the courtyard is important. Projects often include residences at ground level on a quiet street to provide a domestic feeling, since such access is associated with a house, and because such access makes the street livelier. When entering directly from a busy street does not allow enough transition from public to private space, a slight setback and/or differentiation of level such as a stair or plinth may add the requisite privacy. But the rowhouses (or flats in the case of Silodam) discussed here enter directly from a side street (De Opgang, Carnisselande, De Beeklaan, De Muzen, and Silodam). For example, at De Muzen, some rowhouses enter through the interior courtyard, while the rowhouses on the wings (almost 40% of the units) enter from the quiet street through a front garden.

The Netherlands' strong tradition of courtyards includes entrances on both street and courtyard sides. The choice normally depends upon the urban context. The projects where a high proportion of housing is reached from the courtyard are Vrijburcht, De Muzen, and La Grande Cour. At La Grande Cour, one must go through the courtyard to reach rowhouses and all the elevator lobbies for the dwellings. At Vrijburcht, only the work maisonettes enter from the street side, and a few lightwell units have direct ground access directly from the courtyard. At De Muzen, about a third of the rowhouses enter through the courtyard, with the rest entering at ground level from the street.

Various ways to design galleries and decks at De Zilvervloot, Vrijburcht, and De Beeklaan

Gallery and Deck

All of the projects incorporate gallery access, which is an efficient way to meet the Dutch regulations for fire code, light, and cross-ventilation to the residences. The proportion of gallery use varies from to De Opgang, where fully 90% of the dwellings enter from a gallery, most of which circumnavigates an interior courtyard; to De Beeklaan's 81% gallery units; to Carnisselande's 71%; to Vrijburcht, where more than half of the units are reached from the extra-wide deck-gallery; to Silodam, where the short galleries serve only 16% of the

Color enlivens double-loaded corridors at Silodam and La Grande Cour

dwellings. Typical for gallery units is having the gallery windows light the kitchen and/or second bedroom,.

Because of their efficiency, galleries have commonly been used to serve social housing. Thus the appearance of layers of long rows of units has become associated with social housing, and designers are sensitive to alleviating the association.[7] Methods of doing so include reducing the length of the gallery, limiting the number of levels of galleries in a façade, using appealing materials on walls and floors, and using transparent materials, such as glass, for railings.

Corridor or Double-Loaded Hallway

Three of the projects employ double-loaded corridors: Silodam (79%), De Zilvervloot (30%), and La Grande Cour (11%). Silodam is both the densest project in this group and the project with the most double-loaded corridors. The most efficient way of organizing the units is to make back-to-back units or skip-stop units that enter from a double-loaded corridor, which represent fully 60% of the units at Silodam. At De Zilvervloot, with the exception of the skip-stop corridor in De Brug, the double-loaded halls are relatively short—generally extensions of vestibule access areas—and are naturally lit,.

Vestibule or Cluster Hallway

Vestibule access or cluster hallways are important in four projects—De Zilvervloot, La Grande Cour, Carnisselande, and Vrijburcht—and are also used for a small number of units in De Beeklaan (the penthouse and a few corner apartments). Unlike any of the other projects, the 44% of the units that have vestibule access in Vrijburcht are walk-up units. Access to all but two of these lightwell units is by a staircase shared by the six units that are stacked two to a floor. The vestibules in all of the other projects are elevator foyers, including those at De Zilvervloot, with

Bright vestibules at Vrijburcht and De Zilvervloot

37% vestibule units, and La Grande Cour, at 38%. Due to the fire egress requirements, these vestibules have direct access to the exterior and so are commonly naturally lit. The limited size of a vestibule (in contrast to a corridor) restricts the number of dwellings that can have access, thus supporting recognition of neighbors.

Outdoor Space

Outdoor space is considered a normal part of a dwelling in the Netherlands. But it is not enough to provide space: it must be livable territory with sufficient size to place furniture, and sufficient spatial definition to create a private territory. For most complex housing, this is the case.

Box 10.7

Access: Design Principles

1. Consider including at least two or three different types of access in a multi-family project for social and visual variety.
2. Use vertical access to structure your housing project both physically and socially.
3. Use natural light for vertical and horizontal access as much as possible.
4. Consider how many dwellings will be using a vertical access point and try to limit the number to no more than thirty.
5. Use horizontal access as a way of creating neighborhood. Keep the number of dwellings on a horizontal access to three to ten.
6. Vestibule access buildings are especially good at forming a corner.
7. Gallery and corridor buildings can form a wall, a bridge, or a cantilever.
8. In addition to using natural light, employ color to enliven a double-loaded corridor.
9. Widen a gallery to create a deck for outdoor space for a unit or social space for the project.

Garden

There is a wide variety of gardens found in the complex housing projects. At ground level they are mostly associated with rowhouses. At Carnisselande, many have open front gardens and enclosed gardens complete with garden sheds and fence at the rear of the dwelling. At De Muzen, most of the rowhouses also have a small garden area in the front of the unit as well as gardens overlooking the park. The latter are spatially defined as separate from the park through shrubs and a level change. Some of the rowhouses at De Beeklaan

Ground level gardens at De Muzen, an entry plinth at De Opgang, and rooftop terrace at De Zilvervloot

have a very small front garden, but the rear garden normally associated with rowhouses is exchanged for a deck at the back (over the parking). At De Opgang and La Grande Cour, even though there are no gardens associated with the rowhouses, rowhouse entries at both have sufficient space for potted plants.

Above ground roof gardens, decks, and terraces are found in large numbers at De Zilvervloot, La Grande Cour, and Vrijburcht. There are also a few at De Muzen and De Beeklaan. Roof gardens are generally located at the top level, or over a roof where a building steps down. Many rooftop gardens at De Zilvervloot are associated with the penthouses, although the rowhouses and several other units also have roof terraces. La Grande Cour has twenty-four units with rooftop gardens in the periscope towers. These spaces are truly on the roof with only small interior access areas. Fourteen of the dwellings at Vrijburcht have roof decks or terraces, while they are associated with about six dwellings at De Muzen, and two at De Beeklaan.

Silodam has spaces associated with galleries and balconies that incorporate plants that cannot be considered gardens, but the atria on the top level may be used as such.

Balcony

Designed in the VINEX period with a thermal break vis-à-vis the wall, balconies are an important form of outdoor space. Every project includes balconies. Because balconies are considered living spaces, they are usually large enough to contain chairs and possibly a table as well. They may be square (typically needing a vertical support) or rectangular (typically cantilevered or suspended over the façade). Privacy can be a concern if balconies can be viewed from above, or if they are side

Balconies at La Grande Cour and De Opgang

by side. Privacy walls are often employed to separate long balconies. Because balconies are usually located on the sunny south, east and west sides, they create interesting shadow lines on a wall.

In projects such as De Muzen and La Grande Cour, many identical individual balconies are used to create a visual pattern on the façade. At De Opgang the balconies form a screen for the façade. Every flat has unit-wide balconies about 1.5 meters (5 feet) deep, with separating privacy walls and access from the living room and a bedroom. At Silodam such unit-wide, open-air balconies are also found, but here within the building wall. These do not have privacy barriers, so small groups of units share them. In contrast, the balconies at De Zilvervloot are very large, more idiosyncratic, responding to the great many particular circumstances of the different buildings, such as orientation, level, view, and adjacent units. The rowhouse balconies at De Beeklaan that are over the parking are very large, although the flats are without balconies. The balconies at De Muzen are also generous, requiring support either from a side wall or from a column.

Loggias at De Zilvervloot

Loggia

Loggias create outdoor space that is less visible on a façade because they are enclosed by glass doors or windows, and do not extend from the building wall, but are behind it, cut into the building. Thus, on the exterior it can often be difficult to differentiate a loggia from a set of windows. Useful in windy areas and noisy areas

Gallery as outdoor space at Carnisselande

such as a façade facing a busy street, typically they employ glass doors that can be opened to make an outdoor space, or closed to create a more interior one. Loggias are primarily found in the larger structures such as Silodam, La Grande Cour, and De Zilvervloot where wind and noise may be problematic.

Gallery

At several projects galleries function as outdoor space in addition to offering circulation (Vrijburcht, Carnisselande, De Beeklaan, and Silodam). As mentioned previously, the deck galleries at Vrijburcht and De Beeklaan are wider than usual to allow space for tables and/or chairs alongside the passageway. At Carnisselande and Silodam, access galleries, are narrow, but also offer a small space for plants and furniture. Although they are equally wide, the galleries at De Opgang do not seem to function in this way, perhaps because the flats have large balconies.

Box 10.8

Outdoor Space: Design Principles

1. Provide an outdoor space for every dwelling.
2. The outdoor space should have access to sun for at least part of the day.
3. In windy or loud locations, a loggia with soundproof and/or windproof glass walls can be a good solution.
4. When units are directly accessed at the ground plane, a garden can be an excellent living space, especially if at the back of the unit away from the street or entry.
5. If the front of the unit is used to provide outdoor living space, be sure to create sufficient separation from the street, courtyard or gallery, and space for a table and chairs, or at least for two armchairs.
6. If a smooth façade is desired, loggias can be an effective way to provide outdoor space.
7. Balconies are best used on a sunny façade, thus east, south, or west.
8. Consider placement of balconies to assure visual privacy from units adjacent or above. Dividers can differentiate balconies when they are placed next to each other.

Syntactical Structure

Although each building has a unique space syntax gamma diagram, there are shared patterns. For instance, all of the diagrams except Silodam have large fans at the bottom that represent the many spaces that are directly connected to the exterior, whether commercial or residential. Examination of the eight projects reveals three basic organizations above the ground level: (1) relatively independent vertical circulation systems with repetitive fans along individual threads (De Beeklaan, Carnisselande, and De Zilvervloot); (2) two primary, interconnected threads high in the system with repetitive fans across threads (De Muzen and De Opgang—with gallery flats over townhouses); and (3) many interconnected threads with idiosyncratic fans (Vrijburcht, La Grande Cour, and Silodam). In terms of Jane Jacob's concern that density might lead to standardization, none of these projects shows evidence

Gamma diagrams of all eight projects

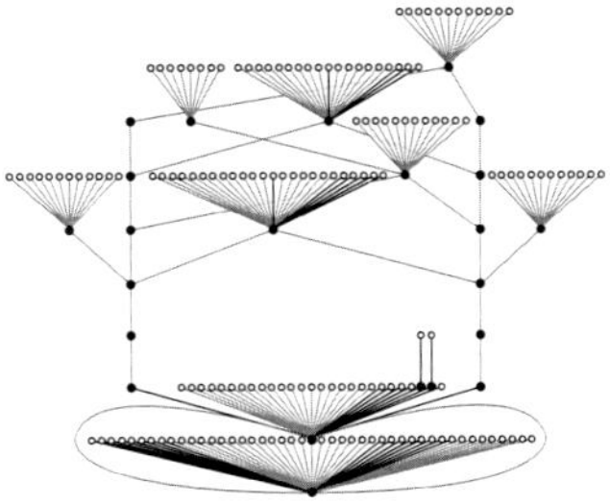

De Muzen

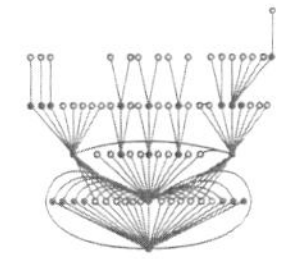

Vrijburcht

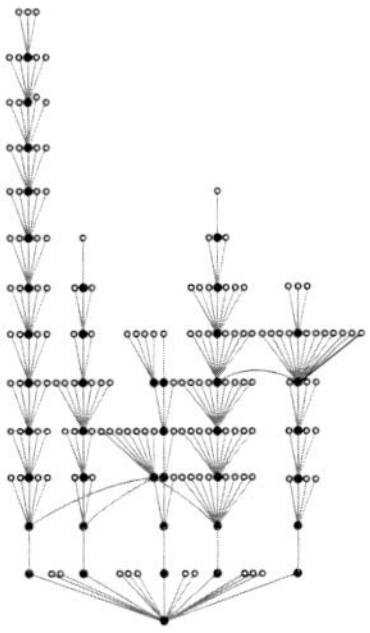

De Zilvervloot

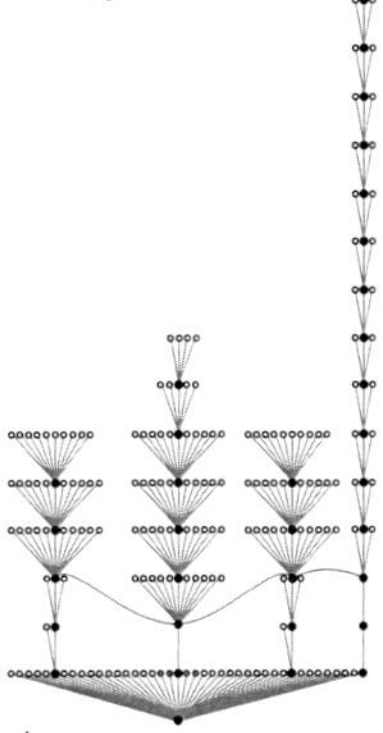

Carnisselande

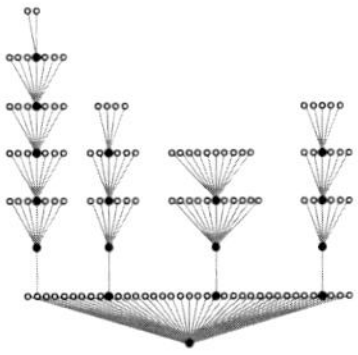

De Beeklaan

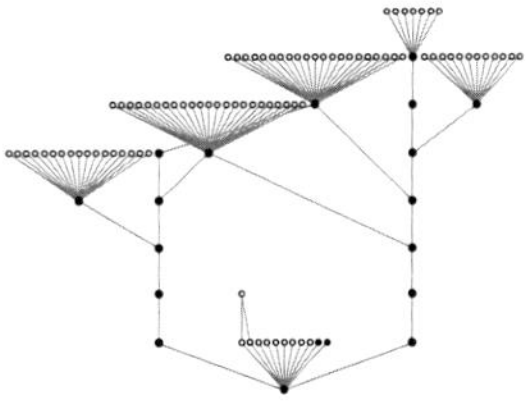

De Opgang

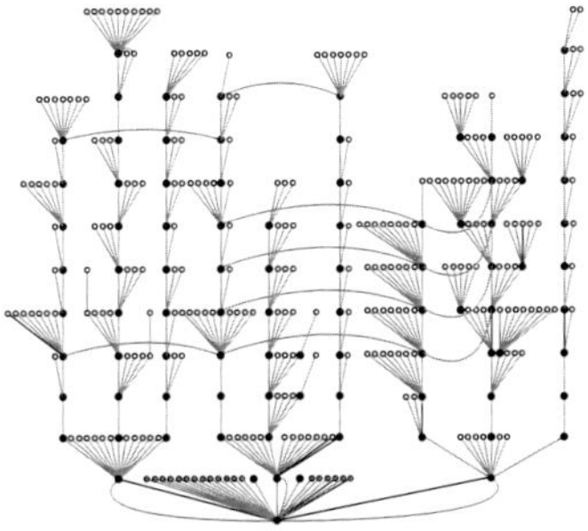

La Grand Cour

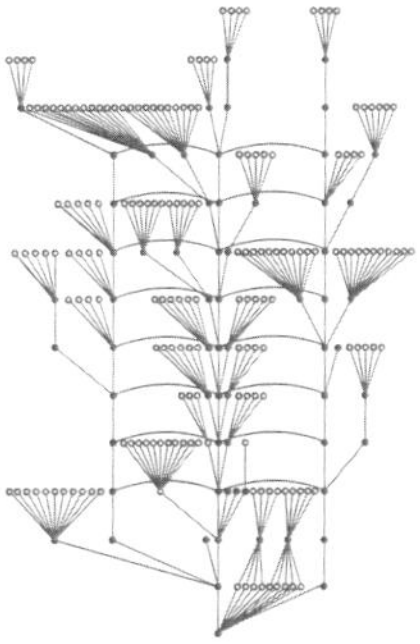

Silodam

Three types of gamma syntax diagrams

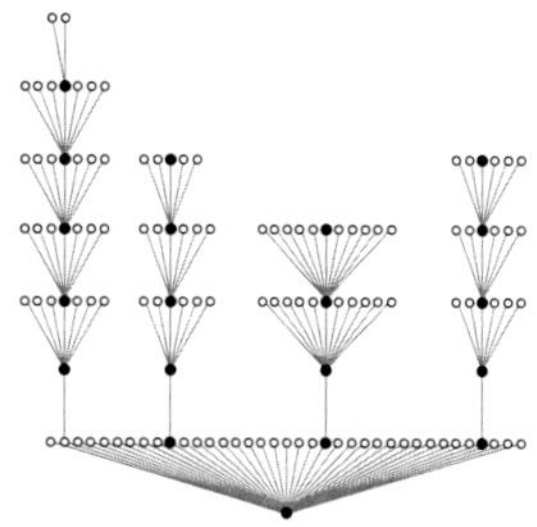

a. Independent threads (De Beeklaan)

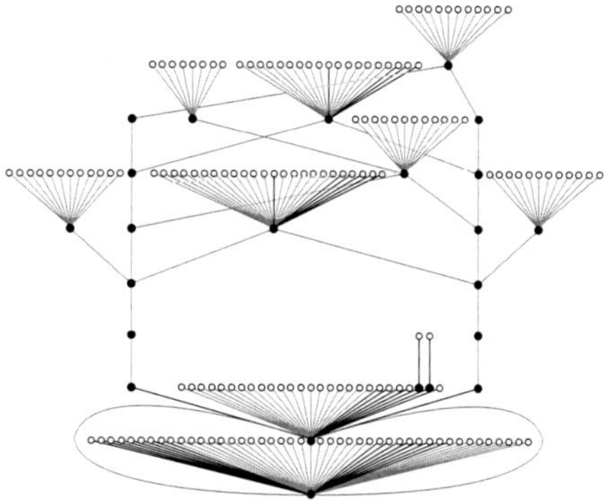

b. Few threads, connected (De Muzen)

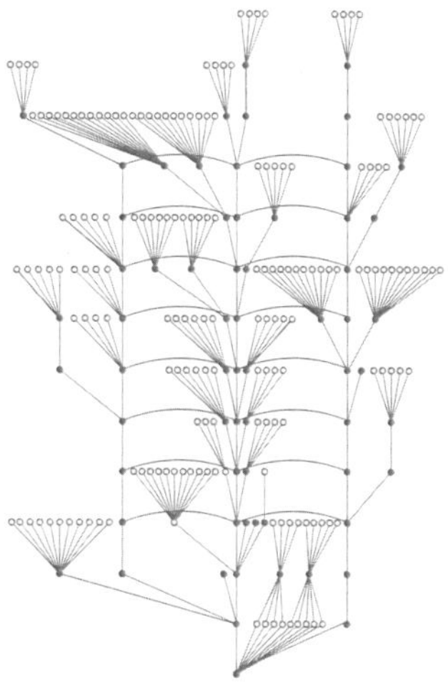

c. Many connected threads (Silodam)

of the kind of repetitive pattern that might cause the monotony she feared. The closest is Silodam that achieves its high density with many double-loaded corridors, yet maintains a great variety of dwelling types.[8]

Overall Organization

Of the three basic types of organization, the projects that are organized around independent circulation represent the most traditional approach to structuring projects. These gather dwellings along the vertical circulation cores as independent buildings. De Beeklaan is the most traditional structure, with four fully independent structures. The vertical cores at Carnisselande and De Zilvervloot, however, are connected at the first floor level so that all the units (in the case of De Zilvervloot) or all the flats (in Carnisselande) can access the communal courtyards and also meet fire exiting requirements. De Zilvervloot has an additional connection at an upper level where the skip-stop units link to two vertical access cores.

Projects that are structured around interconnected vertical circulation are more internally connected than the first set. De Muzen and De Opgang are both a combination of rowhouses and gallery-accessed flats that form a courtyard. Once in an elevator core, one can reach all of the other units in the complex (except the rowhouses at De Opgang).

The third set of projects with many interconnected threads, Vrijburcht, La Grande Cour, and Silodam are more idiosyncratic than the other two types. Vrijburcht is the most highly connected at the bottom both internally and externally, while La Grande Cour and Silodam are more highly connected at upper levels through corridors. At Silodam and at the east wing at La Grande Cour, the corridors are connected between elevator cores, while at La Grande Cour there are also connections at the top, where the cantilevered dwellings connect to two vertical threads for fire exiting.

Entrance

The fan shapes at the bottom of the diagrams represent the spaces that are entered directly from the street. Many of these at De Muzen, Carnisselande, and De Beeklaan are rowhouses. At Carnisselande and De Beeklaan, many are also commercial and/or civic spaces, whereas at De Zilvervloot and La Grande Cour, all of the directly accessible spaces are commercial. The small number of spaces entering directly from the street at De Opgang and Silodam are dwellings and spaces for non-housing activities. At Vrijburcht, streetside entrances do not serve housing but live-work, commercial and community spaces.

The entrances to vertical access cores are direct from the street at De Opgang, Carnisselande, De Beeklaan and De Zilvervloot, while De Muzen, Vrijburcht, and La Grande Cour reach their vertical access cores through a courtyard.

Vertical Access

The number of threads in a project relates to the size of the project and its division into groups of units along a vertical access core. The number of cores or threads varies from two or three at De Muzen, De Opgang, Vrijburcht, and Silodam to four or five at Carnisselande, De Beeklaan, and De Zilvervloot and nine at La Grande Cour (three for each courtyard) (see Table 10.4, page 127). Each of the threads, except those for the lightwell unit stairs at Vrijburcht, represents an elevator that connects a varying number of dwellings. None of the complexes has more than fifty-two units to a core, with De Muzen and Silodam at the maximum with fifty-two and fifty dwellings, respectively. Most commonly projects have between twenty-five and thirty-six dwellings per core, such as De Zilvervloot, De Beeklaan, Carnisselande, De Opgang, and De Beeklaan. Vrijburcht has lowest number of residences per access core at sixteen.

Thread

The character of the threads reveals repetitive or distinct patterns along and between the vertical circulation cores. Carnisselande is an example of regularity across threads, with numerous fans all about the same size, because the number of units on a corridor is quite consistent in the complex. In comparison, the cores at De Zilvervloot are relatively consistent within the threads, but vary between them. Other projects with relatively consistent fans are De Muzen, De Opgang, and Silodam. Vrijburcht and La Grande Cour show the greatest variety both along the threads and between them, indicating a unique character at each level. Of course, those projects with skip-stop maisonettes (De Zilvervloot, La Grande Cour, and Silodam) will have some threads that vary greatly between levels because some levels have no corridors at all.

Fan

The fan-shaped structures along the threads represent how units connect to their access corridors. When discussing access, we addressed the number of units per elevator. The housing fans show how many units are on a hallway or gallery. We find the largest fans at the first group of projects, De Muzen (24) and De Opgang (23), where two elevator cores serve the fans. The largest fans associated with a single elevator core are at Zilvervloot (12), De Beeklaan (11), La Grande Cour (10) and Silodam (10), mostly associated with skip-stop units. But many of the fans are far smaller in size, representing vestibule access or small numbers of units associated with a gallery or hallway—see especially Vrijburcht, De Zilvervloot, and La Grande Cour.

Connection

We have discussed internal connections, showing that De Beeklaan, Carnisselande, and De Zilvervloot are relatively unconnected internally, and that De Muzen,

At Carnisselande each wing serves a different income level

At Silodam each corridor segment serves a different income, but the incomes may be mixed along connecting hallways

At De Zilvervloot, the De Steven tower houses all three income levels

Box 10.9

Syntactical Structure: Design Principles

1. Depending upon the site conditions, have rowhousing, commercial, civic, and/or community space that enters directly from the street to keep it active.
2. Develop vertical circulation cores that either enter directly from the street or from courtyards.
3. Create groupings of units no more than twenty-four per level if there are two sources of vertical access, or twelve if there is only one.
4. There is no need for every level to have the same number of units. This can vary according to the types of units and the site constraints.
5. Consider connecting elevator cores to create opportunities for access to communal space or exiting, taking into account security needs. Creating more opportunities for people within the building to meet fosters awareness of other residents and a sense of community within the complex. However, it is important for security that the access be limited to building inhabitants.
6. If you are interested in having a highly social courtyard, make a circulation path that requires all members to pass through it.

De Opgang, Silodam, Vrijburcht, and La Grande Cour are internally connected. Although all of the complexes have dwellings or non-housing areas that are directly connected to the exterior, only three have a high level of connection of the complex itself to the exterior: De Muzen, Vrijburcht, and La Grande Cour.

Social Principles

A key attribute of complex housing is the mix of incomes, the mix of units for rent and for purchase, and the resultant mix of lifestyles. A designer must be sensitive to creating combinations that will allow different kinds of people to live side by side in social harmony without conflict. Even though a design cannot solve social problems, it can support the creation of community that may prevent them. When people recognize their neighbors, can distinguish neighbors from visitors, and feel committed to the social integrity of their dwelling complex, informal surveillance is very effective security. As addressed earlier, this is affected by the ways units are distributed, the structure of circulation patterns, and whether and how shared amenities are provided and located. In complex housing, the presence of the residents' association that is required by law is also a key factor in developing the necessary community of residents, especially when the developer maintains good contact with the renters.

Mix of Tenure

Having units for sale and purchase in the same complex has the advantage of providing both stability and flexibility for inhabitants. In the cases studied, the actual mix of units for rent and for purchase was decided as much after the units were built as during design. Plans for the number to be rented or purchased, although planned in advance, was unpredictable by developers, especially during the latter part of the time period of these projects, 2008–2010, due to the recession of 2008. During the earlier period, the housing was to be one-third social housing, one-third middle-income housing, and one-third upper-income housing, and at least 33% of the housing was intended to be rental. At the end of the period, the national government sought greater emphasis on home ownership, and housing corporations were encouraged to provide more affordable housing that could be bought. Furthermore, at this later time, due to difficulty in acquiring mortgages, developers had trouble selling the housing. For these reasons, housing designed to be sold ended up being rented and vice versa. In many instances housing originally designed for one income group was rented or sold to a different one. This suggests that there is value in not differentiating between housing for purchase and for sale in the design process.

Mix of Income

The mix of income levels was also somewhat contingent. For example, we saw at Carnisselande that housing originally designed as rental social housing ended up being sold to people of a slightly higher income. Furthermore, the standard units did not differ substantially between those originally designed to be sold and those originally designed to be rented.

On the other hand, there was a conscious approach to handling the mix of units at the different complexes. Some projects mix incomes at the whole complex level, but a given elevator core is to be designed for one income only (Carnisselande and De Beeklaan) (see diagram a, page 240). La Grande Cour mixes two levels of income in an elevator core (see diagram b, page 240). Generally, a core serves either low- and moderate-income units or moderate- and upper-income units. Other complexes maintain a single income at the level of the floor, or syntactical fan, but mix incomes within the elevator core (De Muzen, De Opgang, De Zilvervloot, and Silodam) (see diagram c, page 240). At Vrijburcht, the income levels were designed to be mixed at the courtyard level, with the lightwell units, designed as low-income housing, having their own access. However, these and other units at Vrijburcht were purchased by people with a variety of income levels, some of whom combined two units into one, so income mix exists throughout the project (see diagram d, page 240).

Mix of Lifestyle

The mix of lifestyle is a natural result of the variety of housing and unit types and tenure. We have discussed how every project has a great number of housing and unit types. It is especially interesting to create

Ways of mixing income in complex housing

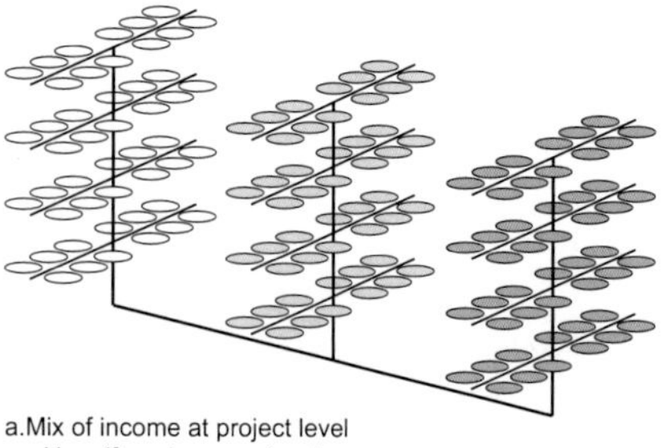

a. Mix of income at project level with uniform incomes in wings

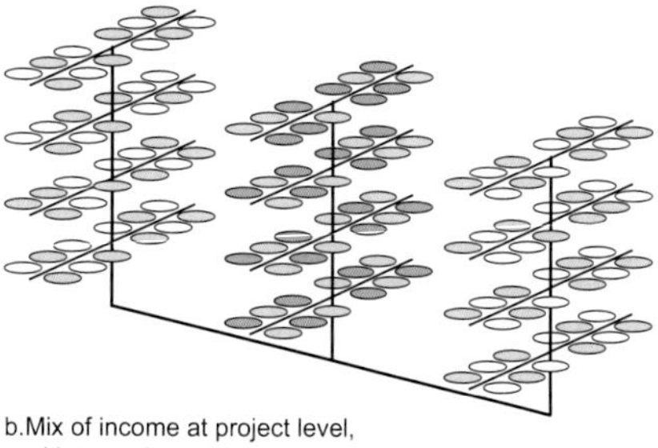

b. Mix of income at project level, with two adjacent income levels in each wing

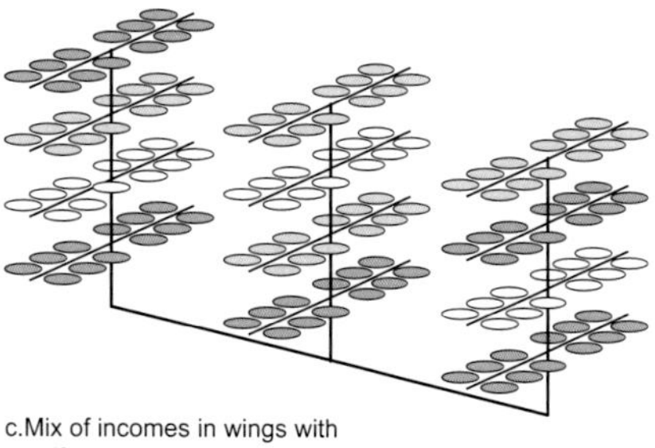

c. Mix of incomes in wings with uniformity in horizontal circulation

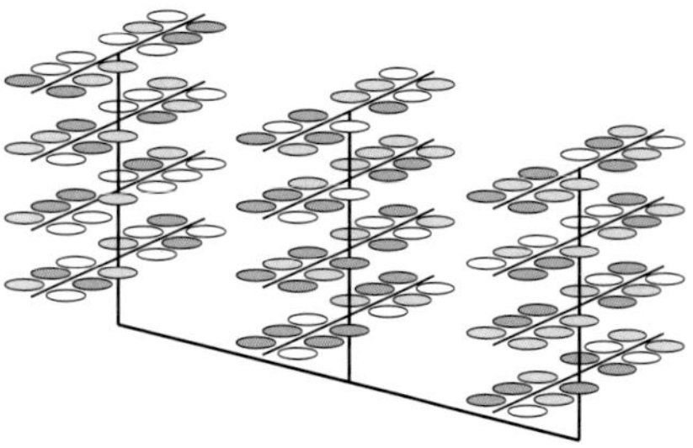

d. Mix of incomes throughout the complex

■ Upper-income ■ Middle-income □ Low-income

opportunities for people of different ages and stages in the life cycle as well as people of diverse abilities. The movement to integrate people of different abilities in housing is evolving to the point where there may not be a need to create special types of dwellings for them, as they will be housed in the same physical dwellings as others, but with social programs that support them.

Community Amenity

We have discussed how the circulation systems of the complex housing projects support the creation of social networks by limiting the number of dwellings on an elevator or corridor. Most of the projects provide space specified as communal, whether a courtyard, terrace, or garden, and other amenities such as mailboxes, meeting space, and bicycle storage. Less frequently found are such things as outdoor seating or communal gardening spaces. Locating these and other community facilities along shared circulation paths creates opportunities for residents to become familiar with each other and to develop social relationships within the complex.

Design Principles for Complex Dutch Housing

The principles set out in this chapter are developed based on the typological analysis of the projects in the previous chapters in order to communicate how one might achieve similar housing, and what might be some of the advantages of doing so. These principles are tied to fundamental, egalitarian choices that value both the individual and the community. At the same time that individual desires and needs are met in the design as a social good and marketing requisite, having an active community—an ingredient essential to the good life—is also necessary to successful complex housing.

The bicycle storage area at Vrijburcht offers opportunities for residents to meet

At La Grande Cour, several community spaces incorporate terraces with views of the city

Courtyards, such as the atrium at De Muzen, are a community amenity

Box 10.10

Social Mix: Design Principles

1. A complex housing design will accommodate low-, moderate-, and upper-income residents.
2. There will be units both for rental and for purchase.
3. There will be a mix of housing and unit types to create a diverse community.
4. Low- and moderate-income units, and housing for sale or purchase, may have identical or similar designs. This allows developers to be flexible in how they meet market circumstances.
5. Combine dwellings of different income levels with care. Units can be:
   a. Mixed without regard to income.
   b. Mixed with similar incomes along a corridor or vestibule, but mixed within a vertical circulation core.
   c. Mixed (by two income levels) in a corridor or elevator core.
   d. Segregated by elevator core, but mixed within the complex.
6. Employ circulation systems, courtyards, and entry designs that support small numbers of residents getting to know each other.
7. Locate community services (mailboxes, communal gardens, meeting rooms, etc.) and include furniture and spaces that support interaction (seating, extra space for conversations) on shared paths to support recognition and interaction among diverse residents.

The following chapter addresses the particular characteristics of the Dutch context that generated complex housing, and what political, social, and professional context might be necessary to implement complex housing both inside the Netherlands, given the changing situation in regard to housing policy, and internationally.

Notes

1. Although Tremblay and Dillman's classic study of housing preferences in the United States is over thirty years old (Kenneth R. Tremblay Jr. and Don A. Dillman, *Beyond the American housing dream: Accommodation to the 1980s* (Lanham, MD: University Press of America, 1983, 45–47)), and housing preferences have very likely become more diverse since it was completed, census figures for 2000 show that 60% of US residents inhabited the single-family home, a number that has been relatively unchanged since 1940 (US Bureau of the Census, "Historical Census of Housing Tables," 2011, www.census.gov/hhes/www/housing/census/historic/units.html).
2. Henk Jan Hollander (developer De Opgang), Interview, The Hague, March 3, 2015, 1.
3. Ibid., 3.
4. As noted in Wijken Dordrecht, *Nieuwsbrief Wielwijk*, January 2016, 3, www.wijkendordrecht.nl/wielwijk/projecten-in-de-wijk/nieuwsbrief-wielwijk-met-alle-plannen-voor-de-wijk and; Gemeente Dordrecht. "HOOFDSTUK 3 Planschrijving nieuwe situatie." ND, www.ruimtelijkeplannen.nl/documents/NL.IMRO.0505.BP155Herzwestervoe-3301/t_NL.IMRO.0505.BP155Herzwestervoe-3301_0002Planschrijvingni.html.
5. Frans De Witte (architect Silodam), Interview, Rotterdam, May 31, 2012, 5.
6. Henk Jan Hollander, Interview, 2012, 9.
7. Fritz van Dongen (National Architect, 2014–15), Interview, Amsterdam, March 2, 2013, 2.
8. In Chapter 1, we noted Jane Jacobs' concern that density might lead to insufficient diversity (*The death and life of great American cities* (New York: Random House, 1961), 209, 211–217).

11 Implications

De Zilvervloot commercial courtyard from above

Complex housing is compelling due to its social mix, relation to context (housing as urban fabric), density, high quality of design, and quality of construction. An outsider looking at Dutch complex housing might wonder how construction of such buildings is affordable. What about the Dutch context makes these possible, and how can these be achieved in other political and cultural contexts? Clearly there are multiple factors that have created the Dutch context. So as to ascertain how this approach can be employed at other times and in other places, this chapter includes voices from interviews with various Dutch housing professionals and addresses both the broad context of the realization of complex housing and more particular aspects of its character.

Design for Density

Because of the scarcity of land and the high population in the Netherlands, the density of housing is taken for granted. Most people seek to own a house with a garden, but typically envision not a freestanding single family house, but a rowhouse, connected to neighbors with a small garden behind. Dwellings are relatively small, and every square centimeter of interior space is taken advantage of.

Among the tradeoffs people make in choosing a place to live, proximity to public transportation often trumps low density in the selection of housing. Density is also important as a strategy to support infrastructure and government services. Automobiles, despite being a primary mode of transportation, are less predominant than in many other places because the Netherlands has developed a strong system of transportation by rail, bus, and bicycle, with people commonly using bicycles to get groceries and take children to school. Additionally, car use is becoming less practical in the western part of the Netherlands as continual traffic jams make automobile transportation slower and more unreliable than public transportation during key travel times. Pedestrian-oriented compact areas are thus very practical. Furthermore, the housing regulations and approach to design practice generate dense housing that is very livable. A modern apartment will have such light, air, interior space, and a decent-sized balcony that is a good substitute for a garden.

Additionally, a certain level of density was mandated by the national government, with municipalities such as Amsterdam choosing to raise it in the design of special urban sites. In many circumstances, higher density offers opportunities for higher profits at the same time that people might prefer to live at a lower density. Designers and developers are thus charged with judicious siting and design of dwellings.

> Now we are in a situation where the consumer is willing to live in cities in high densities. We have to offer good outdoor rooms and architects want to develop nice shapes. (Rein Willems, developer)[1]

Social Mix

When the Dutch national government, seeking balanced neighborhoods, set a goal of 30% social housing for new areas of development, developers provided this balance within their individual projects.[2,3] As we saw in the earlier chapters, housing within each project serves different groups (luxury apartments, penthouses, live-work units, group homes, rowhouses, gallery flats). And designers combined these units in different ways—sometimes similar units and incomes were placed together, at other times unit types and incomes were combined—apparently with equal success.

Although the mix was not without glitches, for example at Carnisselande, the problems that took place appear to be typical of those that might occur in any neighborhood. Additionally, it appears that the mix was unnoticed or unquestioned by residents of most of the projects. Why would this be? One possible explanation is the lack of great income disparity in the Netherlands (the Gini Index that measures disparity on a 100 point scale for the Netherlands is 25, versus 45 for the US, 38 for Japan and 31 for the European Union overall), and the fact that 30% of the Dutch population is eligible for social housing.[4,5] At De Zilvervloot, architect Lucien Kroll noted that the luxury housing was the first to sell in the Dordrecht neighborhood previously exclusively social housing.[6] This may be at least partially attributable to the quality of the dwelling design. But it may also be that in such large projects the mix of residents is seen as normal and in the Dutch context becomes invisible.

At Vrijburcht, perhaps a special case, low-income and disabled residents were actively sought. The future resident design participants and the architect had to make an extra effort to assure that low-income residents would get financing, and agreed that the price of other units would be higher to support a mix of incomes in the project. The housing corporation financed and maintains the group home. The fifty-two-unit size of the project was also thought to be important for the social mix. It was small enough to have a shared community, but large enough to allow economies of scale, such as sharing costs of the low-income units.

As described earlier, the mix of units for rent and for purchase emerged from the government's desire to increase home ownership during the VINEX period, and consequently the government permitted housing corporation developers to sell some of their rental units. Membership in the government-mandated owner associations (created to maintain the projects with mixed tenure) combines homeowners, city and business-owners (if any), and representatives of the housing corporations.

Developers and architects were happy to design and develop housing for a mix of incomes and tenure, with developers having some concern about its viability. But in fact mixing incomes seems to have been non-controversial. Without the government's participation in the planning process, however, it is unlikely the experiment would have taken place, and now it appears to be an accepted practice. As stated by Henk Jan Hollander, developer of De Opgang, "They are still promoting mixing and I don't think that will change."[7]

Role of Quality

An important factor in Dutch having high quality housing is housing corporations.

> We have a history of housing corporations. . . . They will own the building for 20, maybe 50 years. They are interested in how to maintain the facades, how the materials will endure for the long term. In the end it is more profitable. The influence of the housing corporations is significant. They are responsible for close to 50% of the houses.
> (Han van den Born, architect)[8]

The designs, materials, and attention to detail of Dutch housing contributes to its being seen as innovative. How do the Dutch achieve high-quality design in an affordable way?

> It's very much within our culture to try to optimize between costs and design. Depending on the state of the economy, sometimes design is decisive and sometimes the cost. It also depends upon who the client is and also the architect. One thing that is different, we see housing as an element you can . . . design. . . . Housing in our education has always been in the forefront. It's a part of culture here, and a part of what the municipalities and councils prescribe. Especially the bigger cities try to push the level of design. So even if as a developer you were not very interested in making the block of a certain quality you would need to attend to it. Welstand

> committees [city-based legally prescribed, expert, independent aesthetic building committees that advise politicians] are very important.
> (Gerard Comello, developer)[9]

> A high quality of design is very Dutch. All designs are reviewed by "Welstand"—a typical Dutch phenomenon.
> (Noud te Riele, architect)[10]

As to whether the move toward privatization will affect the quality of designs, one developer suggests not.

> It always has to do with money. In the US you sometimes see such a luxurious building that I am jealous. I envy the beautiful materials. We are a much more equal society and that influences the architecture—the social housing always looks nice. As a developer I sell for quality.
> (Henk Jan Hollander, developer)[11]

Housing as Urban Fabric

Depending on the context of comparison, from suburbia to highly dense conurbation, complex housing may seem not dense enough, of normal density, or higher than usual. Nevertheless, it offers a way to design dense housing that is not monolithic and that enhances its neighborhood. When housing is understood as urban fabric rather than as isolated buildings, the planner, urban designer, developer, architect, and landscape architect can build up to the street edge to create courtyards, determine street and sidewalk widths, and locate bike paths. Different types of housing may be combined in a single block along with other functions like stores, cafes, childcare centers, schools, and theaters at ground level or placed on an upper level within a complex. Plants, trees, and gardens can be planted in a courtyard or on extra-wide sidewalks. Designers can balance tall structures or towers with shorter types of housing that allow sunlight to reach all dwellings. They can create patterns for neighborhoods, defining an area entrance or edge with built form, and locating various types of housing and other functions on a busy main street or a quiet side street, at ground level or above. Seeing housing as fabric, where buildings are not necessarily freestanding but connect with the adjacent structures, enables shaping the city in a versatile and free way. According to developer Gerard Comello, in the Netherlands "Even developers, but especially architects consider their building as a part of the urban fabric."[12] The Dutch approach is well summarized in the following observation.

> What we have in Holland is a large tradition that started in the 17th century because all of Amsterdam city has been developed with central planning. We had a tradition of plans and not to just build anywhere you want on a grid. At the beginning of the 20th century Berlage tried to create the newest part of the city with a set of rules related to the middle ages—squares and axes. The structure of the city and the public spaces are much more important than the individual buildings. Public buildings were located in a special spot.
> (Eric Amory, developer)[13]

Thus conceiving of housing as urban fabric depends upon a plan, a vision of neighborhood identity, and of an urban area as a whole. In countries where neighborhoods accrue haphazardly, in response to developers' independent and unrelated proposals, and planning is not done by municipalities, addressing housing as urban fabric is more difficult, if not impossible to achieve. In the Netherlands a municipality's professionally trained municipal urban designers create the structural design for a neighborhood. Local governments then hire developers based on a competitive process, and work with them to implement the urban scheme that may diverge the original plan. The municipality and the developer are both motivated to do this together, quickly, economically, and with style.

Role of Planning and Privatization

> I'm not Dutch—originally, I come from Belgrade. . . . Planning I saw [for the first time] when I came here, and I was very surprised of course. Nobody in the world does it like the Dutch. . . . They decide something, then they do it.
> (Mirjana Milanović, urban designer)[14]

As this book is being written, planning for housing has shifted from the national level to the regional and local levels, as there seems to be a general consensus that national government planning was not as effective at meeting people's needs as the capitalist market process would be, and that home ownership should be encouraged. An outsider wonders, what will the new approach lead to?

For many Dutch developers the market system offers a kind of freedom not available in the previous, highly planned Dutch planned system. They also believe that the free market serves consumer needs better.

> We need a more marketing-oriented development process. It's the government who is in charge of and decides about program. And developers are only allowed to design what is planned by the government, which means the developers in Holland don't have enough influence at the beginning, and only at the end do they have control. . . . I am not worried about the position of the developer; I am worried about the consumer. Consumers are forced to accept a housing system that doesn't meet their demands.
> (Ilta van der Mast, process manager)[15]

> City plans are OK and neighborhood plans are OK, but please give freedom to the developer. Make general rules, but not specific rules. We have so many rules we can't make things.
> (Chris Jagtman, real estate consultant)[16]

Others are not as sanguine, seeing a danger that the new approach will leave some groups unserved, or seeing it as a temporary trend.

> I think it is a sad situation. In Holland we cared a lot about our poor people. The Housing Act of 1901 was a very important breakthrough. Endless speculation was diminished. Rents are rising too hard at this moment. Privatization is a good thing for people with middle incomes. Right now we think that everyone can take care of his own life. . . . I foresee great problems in society.
> (Noud te Riele, architect)[17]

> This liberalization has under the present government . . . an ideological aspect: . . . to suppress the former (social democratic) power of the building corporations, by raising a special tax for building corporations. . . . Maybe under a new government building corporations will get more space again for initiatives.
> (Roelf Steenhuis, architect)[18]

> The market is seen as the best solution for everything. But that is the system that is doing the harm. It will end when the problem it creates will be more expensive than the alternative.
> (David Wolterbeek, developer)[19]

Will the new approach lead to inequities such as homelessness? With the national governmental planning in the 20th and early 21st centuries there has been almost no homelessness in the Netherlands, and when there was, it was immediately addressed.

> For example, we had a very large population of homeless people in the big cities 6–7 years ago. A large program was created to get them off the streets. It was an effective and good cleaning job. They put programs on it. "We give you a normal house for 4 people -and someone will visit you every week", or [a residence] in special buildings . . . Even when housing production is booming, you don't see empty houses. All the houses are occupied.
> (Eric Amory, developer)[20]

My interviewees generally agreed that the change to a market-based system did not mean the end of government participation or planning.

> I [have] noticed that if you leave everything to the market it won't work. I'm very market oriented, but I would never say we don't need government. . . . At the basis, we are still very much public-oriented. . . . Even if we say that we are market-oriented, the direction of the pendulum is to . . . government more of the time than to [the] private.
> (Chris Jagtman, real estate consultant)[21]

> Planning won't be over. Without planning nothing will happen and it's also a way that people will participate. There is a shift in the market. At first we thought it was the crisis that people weren't buying in the suburbs. People are staying in the cities. Everything I do in Amsterdam and The Hague is doing well. The pressure in the housing market in Amsterdam is enormous. But the characteristic of the market that is a bit different. You won't see big developments outside the cities.
> (Henk Jan Hollander, developer)[22]

Citizen Participation

With the apparent exception of La Grande Cour, all the complex housing projects presented here included citizen participation at some level. Such engagement is encouraged by the Dutch urban design procedures that give neighbors, when they exist (unlike La Grande Cour), the right to be consulted and to have their needs met. At Carnisselande, for instance, the existing village retained its identity independent of the new development area; at De Beeklaan, the neighbors' desires for a traditional design were met through negotiations with the city; and at De Zilvervloot, the neighborhood group engaged in a several-year participatory planning process to redesign their part of the city of Dordrecht.

However, Vrijburcht is the only example presented of a Collective and Private Patronage (CPO) project developed by the future residents themselves. As described earlier, the process took five years between the first proposal for the site in IJburg in 2001 and its construction. This process was supported by the city of Amsterdam having set aside land for such projects in the newly created IJburg sandbar. Additionally, the city was instrumental in getting financing for the low-income housing purchasers. According to founding resident Johan Vlug, the role of municipalities in creating such opportunities in the Netherlands is essential, and not all cities provide the support that Amsterdam does.[23] CPO projects, where future residents invest in the development of collective housing, usually require support in order for the future residents to finance design, land purchase, and construction. In this case, the future residents did not get the break in land cost they had anticipated, but nevertheless were able to identify a housing corporation that supplied the capital to continue development.

Even if citizen participation is an accepted practice in the Netherlands, today it primarily consists of neighborhood engagement, unlike the situation in the 1970s and 1980s where residents often actively participated in housing design. Engaging the inhabitants from the beginning has lots of long-term advantages for a project. However, when looked at from a purely cost perspective, its advantages may seem overshadowed by the disadvantages.

> It's nice to do, but it is very complicated, which is why professionals (like developers) don't like to do it.
> (Eric Amory, developer)[24]

In the Netherlands today, specifically in Amsterdam, there is a move toward customer-oriented building and away from the process of government decision-making about housing supply as practiced in the VINEX period. A pioneer in developing Collective Private Commissioning (CPO), architect Hein de Haan describes the process.

> We are moving increasingly from supply-driven to demand-driven building. And that means separate lots and CPO. And those who haven't realized it will find out. Bottom-up development, in other words. An occasional private home

> on a separate lot is nice, but that won't suffice as the basic building block of the city. You end up with a great many low-density suburbs without employment, very few services and almost no public transport. What's more, production per private home is unnecessarily expensive and inconvenient, in contrast to CPO, where a dwelling soon enough comes out under the market price.
> (Hein de Haan, architect)[25]

Another recent example of this demand-driven approach to citizen participation in Amsterdam (and elsewhere in the Netherlands) is called "self-development." In the past, when the city opened up new land areas for development, the large housing corporations have had the exclusive rights to make proposals. But after the recession of 2008, no development was taking place, so the municipality decided to offer undeveloped zones, especially those in "less desirable" areas, to individuals and collectives to build for themselves. This has largely turned out to be a success, with creative low and moderate-income people having access to land at reasonable prices, and in many cases, turning marginal, abandoned industrial areas into residential neighborhoods. However, such developments are biased toward citizens with the wherewithal to build, whether based on economics, motivation or knowledge. Amsterdam urban designer Mirjana Milanović was part of a team that proposed this approach in a former industrial area called Buiksloterham:

> All the artist and alternative types found Buiksloterham interesting. . . . It was very cheap. The building lots were 45,000 euros and now five years later they are 5–6 times the price. There was a reason why people with no money and imagination would buy these lots.
> (Mirjana Milanović, urban designer)

Some of the supports to citizen participation that are offered by municipalities in the Netherlands are financing for low-income residents, creating policies that set aside land for self-development by individual collective groups, and assigning municipal employees to work with neighborhoods. Also potentially supportive for such projects are Dutch housing corporations, and the banks they work with that fund low-income housing and community amenities. Therefore engaging citizens in other countries will likely also require involvement of governments, non-profit organizations, and banks motivated to invest. Citizen participation is valuable because it typically results in projects that fit in their context, are accepted by neighbors, and when future residents are incorporated, fit the needs of inhabitants.

> It's good to involve the people who will live there. Involving the people is always better. Since they are more connected to each other and to the neighborhood, the project is better. They learn to meet each other when the building is planned.
> (Eric Amory, developer)[26]

> Since the quality of design and the maintenance of the buildings are subject to the cultural awareness of the persons concerned, [directly planning and designing with citizens] is the only way, in my opinion.
> (Mieke Bosse, architect)[27]

Perhaps the clinching argument for citizen participation in the design of housing was pinpointed by developer Rein Willems. "The biggest profit is when the consumer is sure to buy."[28]

Role of Professional Designers

The high quality of housing design in the Netherlands is in large part a result of the participation of professionally trained designers. In the Netherlands designers are respected, for as shown before, the population is well aware that it inhabits a designed environment that needs to be cared for and maintained. But in relation to housing, this is also because of H. P. Berlage and others' fundamental influence on the Dutch construction culture at the beginning of the 20th century that made Dutch architects and design professionals part of the construction system in a way not found in many other nations. Developers appreciate what architectural professionals contribute.

> The architects always have another view on things. . . The best projects are [a result of] tension between conceptual design and management. It is not good if either side is too strong. (Rein Willems, developer)[29]

By law architects are involved in all construction projects, and trained design professionals in the planning process. Because in many municipalities developers compete to get commissions and must work with professional urban designers to get their projects built, it is to their advantage to hire urban designers, architects, and landscape architects to create compelling designs and to negotiate with municipalities. Furthermore, most municipalities include a Welstandscommissie (urban aesthetics committee), comprising citizens, architects, and other design professionals, that reviews all projects to assure that they fit their urban setting. Thus aesthetics, tempered by other important considerations, are a normal part of housing construction decisions.

On the other hand, professional designers are not free from criticism.

> Privatization means we have to . . . hire architects who can listen to consumers (instead of competing with colleagues to be in Dutch design books). I am cynical about Dutch architects who don't listen to the real estate market. (Chris Jagtman, real estate consultant)[30]

Role of Developers

The elevated quality of housing is also due to the culture of development in the Netherlands. When the housing corporations were building social housing for rental, they established the pattern of long-term thinking for design and construction that became normal for all Dutch housing. The idea is that social housing should be equal to the housing of the middle class, thus built with good quality materials so it would be easy to maintain. Because they would reap the advantages of a wise investment in rental housing, the associations were motivated to build for the long run.

In all the projects discussed here, developers such as Rein Willems at Bouwfonds for La Grande Cour, David Wolterbeek and Jan de Vletter at Groene Stad Almere, or Eric Amory at De Key for Silodam were committed to building projects well. The developer is in a challenging position of balancing many considerations. Here is a description of the approach taken in developing De Opgang to combine housing and a church.

> As I am an architect and also an engineer myself, I am also involved in the design. The construction here was quite complicated. I have to save money even though it's not the best market in the city—the income is the lowest of the whole city. The more money you save in the construction part of the building the more you can spend on the architectural part. That was really a challenge. We spent a lot of time improving the structure because concrete is not a very architectural material. The officials made a calculation and were surprised at the low cost. The design had to generate income for the church, also income for the city since part of the land was owned by the city. (Henk Jan Hollander, developer)[31]

Now that the Dutch system is becoming marketized, the question is whether developers will be less likely to build high-quality housing. In many other countries, including the United States, housing is an investment commodity that developers may sell to others rather quickly after construction rather than being a place they are committed to for a long period of time. This does not incentivize the developer to build for the long term. But most of the Dutch interviewees did not see this as a problem.

> I don't see the opposition between design and affordability and profit, because you can make beautiful designs and even make a profit. I always say it is only about thinking. If you are too lazy to think, it's true. If you are not too lazy to think it's not true. From my point of view, it has to do with attention.
> (Koen Crabbendam, architect)[32]

Experimentation and Innovation

Since the Dutch are world-renowned for their innovative housing designs, a question that arises when first discovering the move to a market-based system is whether developers, being required to be motivated by profit, would be less interested in innovative design. For some interviewees it was of concern.

> I think there will be less experimentation because of the current economic problems. Companies can't afford to make projects like this [De Beeklan], which is a big problem in the Netherlands right now. There are a lot of suburbs, where it is cheaper. Without money from the government people will not make purchases like this.
> (Wicher Mol, developer)[33]

> The subsidies for experiments used to be many times as large. [Although] there still are areas with experiments, there used to be more, and more interesting examples. Now their experiments are for marketing, whereas before experimentation topics were totally free.
> (Eric Amory, developer)[34]

Only time will tell. The difficulty of relying exclusively on the market to provide housing and to support experimentation has been that developing housing for low-income people is not a profit-making venture. A market-oriented society needs to assure that housing is available to those who need it. This is why the Dutch government created the housing corporations in the first place, and why non-profit housing organizations have come to exist in the United States. As developer David Wolterbeek notes, "Housing corporations were not profit-driven, but socially driven."[35]

Yet most of the interviewees thought that although there might no longer be government support for experimentation, and the amount of experimentation might decrease, the Dutch would always experiment and be housing innovators.

> The experimentation—I do it myself now. As long as it is in the tradition, it will survive. . . . You see very good experiments now. People buy their unit and do it themselves. That's ok—that will be more of the future. The art of the traditional things will be maintained but much more commercial and mixed income.
> (Henk Jan Hollander, developer)[36]

> We still design innovative projects in Holland. It's in our architectural nature. We have little space for a lot of people.
> (Noud te Riele, architect)[37]

> The Dutch like experimenting. To be called conservative is an insult.
> (Ton Schaap, urban designer)[38]

> Partly the experimental will always stay there. But it will become smaller, because people don't want to take a risk. . . . At this moment you see that it would be a market for experimentation. You want to come up with new things . . . not the same thing that's already there. Experimentation will always be needed to move forward.
> (Frans de Witte, architect)[39]

Special Time and Place?

During the VINEX period the Dutch government engaged in national planning; housing corporations were just beginning to create investment capital by selling houses as well as leasing social housing; the economy was booming; and an enormous amount of housing was built. Rental social housing was still seen as an important sector, and the government policy for approximately 30% social housing, and 70% moderate- and upper-income housing was seen as a reasonable goal to achieve an egalitarian society. Designing and developing housing was an important part of the economic engine of the Netherlands.

Also at this time, the city of Amsterdam was eager to develop in new areas and wanted to encourage developers to take the risk. The developer of Silodam acknowledges the city's support.

> Because this was the first project in the [Westerdok] area, we paid a very low amount for the price of the land. [They said,] "This is a city project and we will give you a very good price." Nowadays they would never do that.
> (Eric Amory, developer)[40]

The developers too were in a special situation with the development of housing in high gear nationally. Amory comments on the development goal of the De Key for Silodam: "The purpose was not to earn a lot of money, but to do a long-term investment in a new city part, add new social houses to the property of De Key and lower the risks to sell the market program as fast as possible."

Just five years later, the recession of 2008 put an end to the housing boom, and a new economic reality, with virtually no capital available for housing, caused an almost total cessation of construction for a period of about seven years. Indeed, this does not seem to be a temporary situation, but a great change of direction. According to Frits van Dongen, chief government architect for the Netherlands from 2011–2015, the 20th century was the "Golden Age of Housing", and we are no longer in the golden age.[41]

Economics and the Role of Government

As we have seen, what is seen as affordable at any time and place responds to a complex play of cultural norms, economic circumstances, participants' values, and government regulations.

The Netherlands' Calvinist background, its culture of discussion, and its perception that housing is a human right, create a foundation upon which government officials, developers, and architects are sensitive to the role of housing in creating a decent way of life for their citizens. Eric Amory, developer of Silodam, states "Until 1990 housing was only for social good. Now it is a thing for speculation and investing to create value."[42] Johan Vlug, landscape architect and resident of Vrijburcht, suggested that in contrast to today's view of a house as a commodity, in the past housing and other services were provided as a societal necessity to support a way of life.

> If you look back to the [nineteen] thirties and the fifties and the sixties, most people were working for big companies. So from that time the labor unions, the government, the municipalities were organizing our society so that the laborers could live in good conditions. There were no cars, only a bicycle, or you took the tram. So the whole society was based on that system. And a part of it of course, was social housing. So you have to organize the

facilities to live in a certain place with schools, with jobs, the medical care in the neighborhood. . . . And for me that is the reason that people were renting homes and not buying homes.

[The] municipalities, the housing corporations the big companies were offering quite good homes in quite good neighborhoods. It was quite normal, like in the [United] States, that people were working maybe 20 maybe 30 years, maybe their whole life for one company. That was the deal.

After the second World War, the government and the labor unions, and the big businesses . . . made an agreement that the wages should stay low in the Netherlands; and they did for many, many years. Part of the deal was that the government and the big businesses were offering good homes for reasonable rents. And that worked from, let's say the beginning of the fifties up till the end of the seventies, and then another economic principle was starting up. So up until the seventies, the whole social housing system was an important tool in regulating the availability of affordable housing.

And in the beginning of eighties, big industries were moving out—they started to produce not directly in China, but in Romania or in other cheap countries in Europe, so it was not so normal any more to stay and work for one business. The wages were growing, so there was more money available and they started to buy houses, their homes. And the difference is that it was seen, and it still is, as an economic, profitable object.
(Johan Vlug, resident and landscape architect)[43]

When housing corporations were building housing that they would rent and maintain, there was a clear incentive to build housing that would last for the long term. Now that more housing is being sold, will the housing corporations continue their good building practices for the housing that will no longer belong to them? If they are uncertain as to which housing will be sold or rented, and if the housing corporation continues to serve as members of the residents' group that maintains both rented and sold housing, there remains an incentive to build well.

Furthermore, in the Netherlands, the culture of building housing to last is well entrenched. This is supported by government policies and regulations. All countries have rules and regulations around the construction of housing that are created for the public good. Complex housing arises to a large degree out of such regulations. For example, we have noted how fire codes and rules for "licht en lucht" have resulted in the widespread use of gallery-accessed housing. The impact of such rules is also evident in the Carnisselande project, when the regulations for outdoor space were reduced for a short period and the developer did not include the full balconies originally designed for the low- and moderate-income housing. The government has played a significant role in maintaining quality in the Netherlands.

The big difference is that governmental influence is much bigger here than in America. The concern for quality is one of the things that government has in its working area.
(Rein Willems, developer)[44]

In the Netherlands we hold the idea the government is relevant, and in the 20th century that the professional knows something.
(Ton Schaap, urban designer)[45]

But not all decisions for higher quality are due to government rules. Willems's company, Bouwfonds, thought that having outdoor space made units more marketable and required the architect to include balconies for units that didn't have loggias.[46] When government regulations set minimum standards, what incentivizes a city or a developer to go beyond the minimum rules? In this case it was a marketing decision.

In other cases, going beyond the minimum requirements may have to do with maintaining reputation. One aspect of complex housing is its size and the prominence of the projects in the Dutch context, where they are commonly located in a highly visible location, often iconic for the neighborhood or city. In this situation, the obvious success or failure of the project has motivated a number of municipalities and developers to go beyond what they might have otherwise done. At De Muzen the city was willing to make some concessions to help the competition-winning project succeed, and at De Beeklaan the developer took a loss as a way to enter an urban market new to them. Thus we see that what is actually built is not a direct result of government regulations or economic contingencies, but of some combination of factors including the social attitude and perceived interests of the project participants.

Nevertheless, with increased privatization, rules and regulations will be required to maintain the bar for decent housing, especially for low-income citizens, and are an important contributing factor to achieving it. Thus one aspect of achieving complex housing can be seen as government oversight.

> You still need some comfortable control especially at the lower end of the market, to make sure that housing is available for people that don't make a lot of money. If you don't have some input from the city to control the housing market, the housing prices will double for some of these people.
> (Frans de Witte, architect)[47]

> I think low-income housing will always need involvement of external money and this money should come from the government. There is certainly a risk at the moment that the housing situation of lower income groups gets worse. . . . Social instability will threaten cities and good housing for low-income groups is part of the prevention of this instability.
> (Mieke Bosse, architect)[48]

Another consideration is the role of government planning. We have discussed how planning supports the approach to housing as an urban fabric, the importance to design quality of having professional designers engaged in planning, and how citizens participate in the process in the Netherlands. An important difference between nations and municipalities is the degree to which infrastructure is provided by the government versus the developer. The risk of a developer is significantly reduced when, as typical in the Netherlands, the infrastructure is already in place and a developer is simply constructing architecture. Leaving the planning of infrastructure to the market, as happens in many suburban areas of the United States, subjects the future residents to possible problems such as compatibility with adjacent systems, or reductions in quality for economic reasons.

When done cooperatively between municipalities, developers, and designers, planning not only reduces risk, but offers opportunities for better ideas to emerge. We have also seen how important it is that municipalities, architects, and developers work together and remain flexible. For example, when the building envelope originally agreed upon for La Grande Cour was found insufficient for the required density, the planners at the city of Amsterdam worked with the developers to find the periscope buildings as a solution that would address the needs of both parties. Similarly, at De Muzen the city was willing to change the borders of the park to create a design that better fit the neighborhood. Such cooperation between parties is seen as essential to good urban design.

> In the best Dutch examples there is a cooperation or opposition from both private groups and government on local and national levels. It is never just ONE organization that can address a process which involves so many.
> (Mieke Bosse, architect)[49]

Interviewees seemed to agree that to make privatization really work, planning cannot be abandoned.

> The danger is that the developer just does what they are really good at and stamp it [out] making copies of it

> everywhere you look. Some planning will always be needed. In the last 20 years' developers did their stamp and built their same houses. They don't contribute to a community feeling.
> (Frans de Witte, architect)[50]

> Now we will build social housing, but not in the best spots. Maybe the cheaper sites. Years ago many different kinds of people lived in social housing, but today only the lowest income. . . . Official figures show a problem.
> (David Wolterbeek, developer)[51]

It may be that if problems arise from privatization, because of the Dutch ethic of caring for the needy, the system of providing housing will be self-correcting.

> I think the system will correct itself if problems become larger. We have a good tradition in that too [in] the Social Housing Act of 1901.
> (Noud te Riele, architect)[52]

Conclusions

In the Netherlands the privatization of the housing market is largely seen more as an evolution than a radical change of direction The more market-oriented approach to developing housing, and limiting planning to the region and municipality, when looked at from the perspective of Dutch architects and planners, is seen as a needed correction to government policy that was out of touch with consumers' needs. Although some voice concern that certain groups' housing needs will not be met, others see the Dutch tradition of caring for those in need as continuing and balancing out such problems in the long run. While national planning has halted, sufficient planning continues to take place at the regional and local level where there is more direct contact with the situation at hand. Most people seem confident that government regulation will limit the potential negative effects of the profit motive.

The special circumstances that led to the design and construction of complex housing in the Netherlands between 1990 and 2010 do not have to be replicated in order to create complex housing at other times and places. However, examination of the impact of different factors reveals several that are important contributors. It is more likely to happen when housing is understood to be urban fabric, when the site is encompassed as part of the larger neighborhood, and when cities have urban designers that understand how to make a physical neighborhood plan. Complex housing is more likely to work in a prominent site on an open space of some kind. It is more likely to be built in response to building codes that stress, among other factors, the importance of "licht en lucht," outdoor space for all dwellings, and building on or close to the lot line. It is more likely to be accepted when citizens and future inhabitants are involved in its planning and design. It is more likely to work when the number of residents that share an entry, corridor, or courtyard is small enough for people to recognize each other as a cohesive group.

To enable the construction of complex housing, municipalities will need to require such things as a mix of incomes and tenure, and to have in place non-profit organizations and banks that will finance low-income dwellings, cooperative housing, and community amenities. Also in place should be mechanisms that allow citizen participation in decision-making. Finally, complex housing requires a municipality motivated to build an urban community with housing for diverse residents, an architect with strong design skills and a developer committed to long-term investment in and long-term oversight of a project.

Notes

1. Rein Willems (Bouwfonds, developer, La Grande Cour), Interview, Amsterdam, March 25, 2010, 3.

2. Eric Amory (De Key, developer, Silodam), Interview, Amsterdam, May 25, 2012, 3.
3. Gerard Comello (Lingotto, developer, Whale) Interview, Amsterdam March 5, 2015, 1.
4. Assuming that Gini changes gradually over time, these statistics are based on information from different years for each place: 2013 for the Netherlands, 2007 for the United States, 2011 for Japan, and a 2014 estimate for the European Union. Taken from the *World Fact Book* (Central Intelligence Agency (US), Website, August 18, 2016, www.cia.gov/library/publications/the-world-factbook/rankorder/2172rank.html).
5. Falk, Nicholas, "New communities—Looking and learning from Dutch experience," *Town and Country Planning* (December 2008), 503–508.
6. Lucien Kroll, (AUAI, architect, De Zilvervloot), Site Visit, Dordrecht, March 2, 2010.
7. Henk Jan Hollander (NL—Development BV, developer, De Opgang), Interview, The Hague, March 3, 2015, 2.
8. Han van den Born (KCAP, architect, De Opgang), Interview, Rotterdam, May 24, 2012, 8.
9. Gerard Comello, Interview, 2015, 7.
10. Noud te Riele (Atelier Pro, architect, De Muzen), Interview, The Hague, March 4, 2015, 7.
11. Henk Jan Hollander, Interview, 2015, 11.
12. Gerard Comello, Interview, 2015, 7.
13. Eric Amory, Interview, 2012, 8.
14. Mirjana Milanović, (urban planner and designer, Amsterdam), Interview, June 11, 2016, 1.
15. Ilta van der Mast (Boulevard Management en Advies BV), Interview, Rotterdam, March 3, 2010, 1.
16. Chris Jagtman (Real Estate Consultant, Stadlicht BV) Interview, Arnhem, March 10, 2010, 1.
17. Noud te Riele, Interview 2015, 8.
18. Roelf Steenhuis (Steenhuis Bukman, architect, Carnisselande), Interview addendum by email, June 28, 2016, 2.
19. David Wolterbeek (Groene Stad Almere/Allantie, developer De Muzen), Interview, February 25, 2015, 6.
20. Eric Amory, Interview 2010, 9.
21. Chris Jagtman, Interview, 2010, 1.
22. Henk Jan Hollander, Interview, 2015, 2.
23. Johan Vlug (landscape architect, resident, Vrijburcht), Interview, June 10, 2016, 16.
24. Eric Amory, Interview 2010, 10.
25. Hein de Haan (CASA, architect, Vrijburcht), in Joosje van Geest, "A generation that is growing up that can't even share a single facility: Interview with Hein de Haan, CASA" *Dash #8*, March 1, 2013, 48–57, 52.
26. Eric Amory, Interview 2010, 10.
27. Mieke Bosse (SCALA, architect, De Beeklaan), Interview addendum by email, September 1, 2016, 10.
28. Rein Willems, Interview, 2010, 3.
29. Ibid., 2010, 2.
30. Chris Jagtman, Interview, 2010, 1.
31. Henk Jan Hollander, Interview, 2015, 4.
32. Koen Crabbendam (CASA, architect, Vrijburcht), Interview, June 13, 2016.
33. Wicher Mol (Bouwfonds, developer, De Beeklaan), Interview, May 12, 2012, 7–8.
34. Eric Amory, Interview, 2010, 1.
35. David Wolterbeek, Interview, 2015, 8.
36. Henk Jan Hollander, Interview, 2015, 2.
37. Noud te Riele, Interview, 2015, 8.
38. Ton Schaap (City of Amsterdam, urban designer), Interview, Amsterdam, March 23, 2010, 2.
39. Frans de Witte, (MVRDV, architect, Silodam), Interview, Rotterdam, May 31, 2012, 9.
40. Eric Amory, interview, 2010, 3.
41. Frits van Dongen (Van Dongen-Koschuch, architect), Interview, Amsterdam, March 2, 2015, 7.
42. Eric Amory, Interview, 2012, 9.
43. Johan Vlug, Interview, 2016, 13.
44. Rein Willems, Interview, 2010, 3.
45. Ton Schaap, Interview, 2010, 4.
46. Rein Willems Interview, 2010, 2.
47. Frans de Witte, Interview 2012, 9.
48. Mieke Bosse, Interview addendum, 2016, 10.
49. Ibid., 10.
50. Frans de Witte, Interview, 2012, 9–10.
51. David Wolterbeek, Interview, 2015, 5.
52. Noud te Riele, Interview, 2015, 8.

Bibliography

AAUI (Atelier d'Architecture, d'Urbanism et d'Informatique). *Selected plans, working drawings and other documentation for De Zilvervloot*. Brussels, Belgium, 2009–2015.

Aedes Vereniging van Woningcorporaties. *Dutch social housing in a nutshell*. Hilverson: Aedes, 2003. www.aedes.nl/binaries/downloads/vereniging/about-aedes/20130807-dutch-social-housing-in-a-nutshell.pdf

Alberti, Leon B. *The architecture of Leon Batista Alberti in ten books*. London: Edward Owen, 1755.

Architectuur NL. *Documentatie: Woningbouw en kerk: De Opgang*. Amsterdam: KCAP Architects&Planners, 2010. www.architectuur.nl/project/de-opgang-amsterdam/

ARCspace. "Silodam (Housing Silo) MVRDV." Website. Amsterdam: ARCspace.com, February 7, 2005, updated December 19, 2013.

Aubuchon, Vaughn. "Vaughns Countries of Europe Summary." Vaughn's Summaries. www.vaughns-1-pagers.com/geography/european-countries.htm

Berghauser-Pont, Meta, and Per Haupt. *Spacematrix: Space, density and urban form*. Rotterdam: NAi Publishers, 2010.

Boeijenga, Jelte, and Jeroen Mensink. *Vinex Atlas*. Rotterdam: 010 Publishers. 2008.

Centraal Bureau voor de Statistiek (CBS). "Netherlands: 2012–13." Figures and Statistics on Wielwijk District of Dordrecht. www.oozo.nl/cijfers/dordrecht/wielwijk

Centraal Bureau voor de Statistiek (CBS). "Netherlands: 2015." Voorraad woningen; mutaties en bouw, 1921–2011." Housing stock: Construction, 1921–2011. http://statline.cbs.nl/Statweb/publication/?VW=T&DM=SLNL&PA=71527NED&D1=5-7&D2=a&HD=151230-1204&HDR=T&STB=G1; http://statline.cbs.nl/Statweb/publication/?VW=T&DM=SLNL&PA=81955NED&D1=0–2,7&D2=1&D3=0&D4=16,33,50&HD=151230–1206&HDR=G1,T&STB=G2,G3

Central Intelligence Agency (US). *World Fact Book*. Website, August 18, 2016. www.cia.gov/library/publications/the-world-factbook/rankorder/2172rank.html

Cervero, Robert, and Erick Guerra. "Urban densities and transit: A multi-dimensional perspective." Working Paper UCB-ITS-VWP-2011-6, Berkeley CA: Institute of Transportation Studies, 2011.

City Cour Combination. *La Grande Cour: Wonen tussen het IJ en de binnenstad* (real estate brochure). Haarlem, NL: Bouwfonds Mab Ontwikkeling, 2007.

Clark, Roger H., and Michael Pause. *Analysis of precedent: An investigation of elements, relationships and ordering ideas, in the work of eight architects*. Raleigh: Student Publication of the School of Design, North Carolina State University, 1979.

De Allantie ALM-XML. "Aangeboden door de Allantie ALM–XML."

De Haan, Hein. "Collective Client Controlled Development of Space: Examples from an Amsterdam Practice." In *Making room for people: Choice, voice and livability in residential places*, edited by Lei Wu and Evert Hasselaar, 129–152. Amsterdam: Techne Press, 2011.

De Vries, Nathalie. "MVRDV brought different house types together with Silodam block says Nathalie de Vries." Online video clip: YouTube. Dezeen, August 10, 2015. / www.youtube.com/watch?v=FmEufmB8B6E.

De Witte, Frans. (Architect MVRDV), Interview, Amsterdam, May 31, 2012.

Doevendans, Kees, and Richard Stolzenburg. *Stad en samenleving*. Groningen: Martinus Nijhoff, 2000.

Durand, Jean-Nicolas-Louis. *Recueil et parallèle des édifices de tout genre, anciens et modernes, remarquables par leur beauté, par leur grandeur ou par leur singularité, et dessinés sur une même echelle*. Paris: Chez La auteur, 1801.

Ebner, Peter. *Typology+: Innovative residential architecture*. Basel: Birkhäuser, 2010.

Falk, Nicholas. "New communities—Looking and learning from Dutch experience." *Town and Country Planning* (December 2008): 503–508.

Ferré, Albert, Tihamér H. Salji, and Actar, *Total housing: Alternatives to urban sprawl*. Barcelona: Actar, 2010.

Galle, Maaike, and Ettjen Modderman. "VINEX: National spatial planning policy in the Netherlands during the nineties." *Netherlands Journal of Housing and the Built Environment* 12, no. 1 (1997): 9–35.

Gash, Notta, and Farkas Chen. "Silodam Amsteram, NL-MVRDV (sic)." Website: silodamvrdvblogspot.com. Jerusalem: Bezalel Academy of Art and Design. Friday November 16, 2007.

Gemeente Amsterdam. "Gebouwd in Amsterdam: De Opgang." ND. www.gebouwdin.amsterdam.nl/main

Gemeente Dordrecht. "HOOFDSTUK 3 Planschrijving nieuwe situatie." ND. www.ruimtelijkeplannen.nl/documents/NL.IMRO.0505.BP155Herzwestervoe-3301/t_NL.IMRO.0505.BP155Herzwestervoe-3301_0002Planschrijvingni.html

Grinberg, Donald. *Housing in the Netherlands: 1900–1940*. Culemborg: Delft University Press. 1982.

Groenendijk, Paul. "Zilvervloot Dordrecht." www.architectuur.nl/107480/Een-project-uitgebreid/De-Zilvervloot-Dordrecht. May 25, 2007.

Groepswonen Door Ouderen. Website www.gdo-denhaag.nl. 3301/t_NL.IMRO.0505.BP155Herzwestervoe-3301_0002Planschrijvingni.html.

Hagens, Rob. "Pourquoi Wielwijk?" unpublished PowerPoint presentation, 2008.

Hanson, Julienne. *Decoding homes and houses*. Cambridge: Cambridge University Press, 2003.

Hillier, Bill, and Julienne Hanson. *The social logic of space*. Cambridge: Cambridge University Press, 1984.

Hillier, Bill, Adrian Leaman, Paul Stansall, and Michael Bedford. "Space syntax." *Environment and Planning B: Planning and Design* 3, no. 2 (1976): 147–185.

Hoekstra en van Eck. *Project De Opgang: Betaalbaar wonen in een bijzonder project in Osdorp. [Real Estate Brochure]*. Amsterdam: Hoekstra en Van Eck Makelaars, 2008.

Ibelings, Hans, and Egbert Koster. "De Muzen Almere, 1993–1997." In *Atelier Pro*, 44–49. Rotterdam: 010 Publishers, 2001.

Jacobs, Jane. *The death and life of great American cities*. New York: Random House, 1961.

Janssen, Rainer, and Nico Jansen. "Vrijburcht: Amsterdam–Oost." Vitale Knoppunten van Zachte Infrastructure, Berlin and Amsterdam, 2013. www.knooppuntenvanzachteinfrastructuur.nl/portfolio/vrijburcht/

KCAP. *83 Appartementen en een Kerk te Osdorp, Amsterdam, Schetsboekje*. Rotterdam: KCAP, 2006.

KCAP Architects&Planners. *Selected plans and diagrams for De Opgang*. Rotterdam, Netherlands, 2011–2016.

Kendall, Stephan, and Jonathan Teicher. *Residential open building*. London: E&FN Spon, 2000.

Komossa, Susanne. *The Dutch urban block and the public realm: Models, rules and ideals*. Nijmegan: Vantitilt Publishers, 2010.

Komossa, Susanne, Han Meyer, Max Risselada, Sabien Thomaes, and Nynke Jutten. *Atlas of the Dutch urban block*. Bussum: THOTH, 2005.

Krampen, Martin. *Meaning in the urban environment Pion*. London: Pion, 1979.

Kroll, Lucien. 2006. "Atelier d'Architecture, d'Urbanism et de l'Informatique: Zilvervloot." *A+U: Architecture and Urbanism* 1006: 6, #429, 82–88.

Laurens, Marianne. "Wonen in een Vinexwijk." *Via Vestia* no. 1: Rotterdam (Summer, 2011): 18.

Markus, Thomas A. *Buildings & power: Freedom and control in the origin of modern building types*. London: Routledge, 1993.

Meyer en Van Schooten Architecten. " 'La Grande Cour' Wooncomplex Westerdokseiland" (unpublished monograph). Amsterdam: Meyer en Van Schooten Architecten, 2008.

Ministry of Housing, Spatial Planning and the Environment. *Summary national spatial strategy: Creating space for development*. Amsterdam: Ministerie van Infrastructuur en Milieu, 2010.

Mitchell, Robert Edward. "Some social implications of housing density." *American Sociological Review* 36 (February 1971): 18–29.

Moudon, Anne Vernez. *Built for change: Neighborhood architecture in San Francisco*. Cambridge: MIT Press, 1986.

Mozas, Javier. "This Is Hybrid." In *This is hybrid: An analysis of mixed-use buildings*, edited by Aurora Fernandez Per, Javier Mozas and Javier Arpa, 12–45. Vitoria-Gasteiz: a+t, 2011.

Mozas, Javier, and Aurora Fernández Per. *Densidad: nueva vivienda colectiva [Density: New collective housing]*. Vitoria-Gasteiz, Spain: a+t Ediciones, 2006.

Mozas, Javier, and Aurora Fernández Per. "MVRDV: Silodam, Amsterdam NL." In *Densidad: nueva vivienda colectiva [Density: New collective housing]*, 382–389. Vitoria-Gasteiz, Spain: a+t Ediciones, 2006.

Municipality of Almere. "The City of Almere." Dutch Website Municipality of Almere. http://english.almere.nl/the-city-of-almere/. (Accessed July 13, 2016).

MVRDV. "Silodam." Website. mvrdv.nl/projects/silodam

MVSA Architects. Selected plans, working drawings and other documentation for La Grande Cour, 2010–2016.

MVSA Architects. Website, 2016. www.mvsa-architects.com/project/detail/29/la-grande-cour

Palladio, Andrea. *The four books of architecture*, Edited by Adolf K. Placzek. New York: Dover, 1965.

Penn, Alan, and Alasdair Turner. "Space Syntax Based Agent Simulation." In *Pedestrian and evacuation dynamics*, edited by Michael Schreckenberg and Som Deo Sharma, 99–114. Berlin, Germany: Springer-Verlag, 2002.

Peponis, John, Catherine Ross, and Mahbub Rashid. "The structure of urban space, movement and co-presence: The case of Atlanta." *Geoforum* 28, no. 3 (1997): 341–358.

Ploeger, Hendrick, Aart ven Velten, and Jaap Zevenbergen. *Real property law and procedure in the European Union: Report for the Netherlands*. Delft: OTB Research Institute, 2005.

Polano, Sergio. *Hendrik Petrus Berlage: Complete works*. New York: Rizzoli, 1989.

Pollio, Vitruvius. *Vitruvius: The ten books on architecture*. Translated by Morris Hicky Morgan. New York: Dover, 1960.

Preziosi, Donald. *The semiotics of the built environment: An introduction to architectonic analysis*. Bloomington: Indiana University Press, 1980.

Rapoport, Amos. "Toward a redefinition of density." *Environment and Behavior* 7, no. 2 (1975): 133–158.

Rapoport, Amos. "Defining Vernacular Design." In *Vernacular architecture: Paradigms of environmental response*, edited by Mete Turan, 68–102. Brookfield, VT: Avebury Press, 1990.

Reh, Wouter, Clemens Steenbergen, and Diederik Aten. *Sea of land: The polders as an atlas of Dutch landscape architecture*. Wormer: Stichting Uitgeverij Noord-Holland, 2007.

Robinson, Julia W. "Dutch Complex Housing: Design for Density." In *Future of architectural research: Proceedings of the Architectural Research Centers Consortium (ARCC) 2015 Conference*, edited by Ajila Aksamija, John Haymaker and Abbas.

Robinson, Julia W. *Institution and home: Architecture as a cultural medium*. Delft: Techne Press, 2006.

Robinson, Julia W., and Mandana Motamed. "Comparative analysis of housing from the Netherlands and the United States" (Unpublished Paper Presented at the International Association of People-environment Studies (IAPS), Lund, Sweden, June, 2016).

SCALA Architecten. *Selected plans, working drawings and other documentation for De Beeklaan*. The Hague, Netherlands, 2007–2016.

SCALA Website. *Woningen en appartementen Beeklaan*. Den Haag. www.scala-architecten.nl/project/woningen-en-appartementen-beeklaan-den-haag

SCALA Website. *Woningen Regent Valkenbos*. Den Haag. www.scala-architecten.nl/project/regentes-valkenbos

Schaap, Ton. *Amsterdam*. Rotterdam: 010 Publishers, 2008.

Sherwood, Roger. "Housing Prototypes: Silodam." Website, 2002. housingprototypes.org.

Steenhuis, Bukman Architecten. "Carnisselande, Barendrecht." Website, 2007. www.steenhuisbukman.nl/?id=3&lang=nl&offset=2&group=16&tag=0&item=0197&view=gallery&image=672, 2007.

Steenhuis, Roelf. "Typologie en plek (Typology and place): Steenhuis Bukman Architecten" (Class Presentation Netherlands Student Trip, Utrecht, May 25, 2007).

Stieber, Nancy. *Design and society in Amsterdam: Reconfiguring urban order and identity, 1900–1920*. Chicago: University of Chicago Press, 1998.

Stokols, Daniel. "On the distinction between density and crowding: Some implications for future research." *Psychological Review* 79, no. 3 (1972): 275–277.

Stolzenburg, Richard. Class Lectures (1999–2004), based in part on Doevendans and Stolzenburg, 2000.

Student research comparing Dutch and US Housing projects, Architecture 5750: Contemporary Dutch Housing in Urbanism (Unpublished Papers, Minneapolis: University of Minnesota, 2007–2010).

Tremblay, Kenneth R., Jr, and Don A. Dillman. *Beyond the American housing dream: Accommodation to the 1980s*. Lanham, MD: University Press of America, 1983, 45–47.

US Bureau of the Census. "Historical Census of Housing Tables." 2011. www.census.gov/hhes/www/housing/census/historic/units.html.

Uytenhaak, Rudy, and Jeroen Mensink. "Westerdoksdijk Amsterdam." *Archimed*, April, 2009. www.archined.nl/en/reviews/engels/westerdoksdijk-amsterdam/

Van der Lans, Jos. *Wonen in Almere 1.0: Almere als dagelijkse evaring*. Almere: Department of Urban Development (Dienst stedelijke ontwikkeling van de gemeente Almere), 2009.

Van Geest, Joosje. "A generation that is growing up that can't even share a single facility: Interview with Hein de Haan, CASA." *Dash #8* (March 1, 2013): 48–57.

Van Heelsum, Anja. *Case study on housing: Amsterdam, Netherlands*. Dublin: European Foundation for the Improvement of Living and Working Conditions, 2007, 20.

Vergunst, Menno, and Johan Vlug. "Vrijburcht." In *CoHousing cultures: Handbook for self-organized, community-oriented and sustainable housing*, 164–179. Berlin: Jovis Verlag, 2012, 169.

VLUGP. "Vrijburcht: Collective and Private Patronage-Also for Public Space." *VLUGP stedebouw & lanschaparchitectuur*, 2–5, Amsterdam, 2010, 2. phttp://docplayer.net/47989657-Vrijburcht-vlugp-collective-and-private-patronage-also-for-public-space.html

Wijken Dordrecht. "Nieuwsbrief Wielwijk." January 2016, 3. www.wijkendordrecht.nl/wielwijk/projecten-in-de-wijk/nieuwsbrief-wielwijk-met-alle-plannen-voor-de-wijk

Wikipedia. "Population of the Netherlands 2006." Website: *Demographics of the Netherlands*, 2016. https://en.wikipedia.org/wiki/Demographics_of_the_Netherlands

Woonbron and Gemeente Dordrecht. "Wielwijk sterk en weerbaar: Visie Wielwijk 2015 en uitvoeringsprogramma 2006–2015." 2007. www.ruimtevoorklimaat.nl/cases/04-Herinrichting-Wielwijk

Woondrecht Vastgoed, Ooms Drechtsteden Makelaars, and Waltman & Company. *Zilvervloot: A living dream*. Dordrecht: Real Estate Sales Document, 2005.

Woondroom Zorg. *De Roef*. Utrecht: Woondroom, 2014. woondroomzorg.nl/samenwerkingspartners/deroef/tabid/397/default.aspx

Yin, Robert K. *Case study research: Design and methods* [Second Edition]. Thousand Oaks, CA: Sage, 1994.

Index

Page numbers for figures, photographs, and tables are in italics.